Palgrave Macmillan Studies in Banking and Financial Institutions

Series Editor
Philip Molyneux, Bangor University, Bangor, UK

The Palgrave Macmillan Studies in Banking and Financial Institutions series is international in orientation and includes studies of banking systems in particular countries or regions as well as contemporary themes such as Islamic Banking, Financial Exclusion, Mergers and Acquisitions, Risk Management, and IT in Banking. The books focus on research and practice and include up to date and innovative studies that cover issues which impact banking systems globally.

Chrysovalantis Gaganis · Fotios Pasiouras ·
Menelaos Tasiou · Constantin Zopounidis
Editors

Sustainable Finance and ESG

Risk, Management, Regulations, and Implications for Financial Institutions

palgrave
macmillan

Editors
Chrysovalantis Gaganis
Department of Economics
University of Crete
Rethymno, Greece

Menelaos Tasiou
School of Accounting, Economics and Finance
University of Portsmouth
Portsmouth, UK

Fotios Pasiouras
Montpellier Business School
Montpellier, France

Constantin Zopounidis
School of Production Engineering and Management
Technical University of Crete
Chania, Greece

Audencia Business School
Nantes, France

ISSN 2523-336X ISSN 2523-3378 (electronic)
Palgrave Macmillan Studies in Banking and Financial Institutions
ISBN 978-3-031-24282-3 ISBN 978-3-031-24283-0 (eBook)
https://doi.org/10.1007/978-3-031-24283-0

Cover credit: QQ7/iStock/Getty Images Plus

This Palgrave Macmillan imprint is published by the registered company Springer Nature Switzerland AG
The registered company address is: Gewerbestrasse 11, 6330 Cham, Switzerland

Preface

Climate change and enhanced public awareness about environmental and social issues have brought important differences to the way of doing business. Sustainable finance along with Environmental, Social, and Governance (ESG) aspects and their implications for financial institutions have attracted the attention of academics, consumers and market participants, and have gained priority in the agenda of policy makers. The financial risks of climate change are becoming of paramount importance for banks and insurers as regulators introduce policies that supporting a low-carbon or net-zero agenda and set out expectations for climate risk stress testing. Yet a lot remains to be done toward this direction. For example, the July 2022 report of the European Central Bank (ECB) in the main findings of its climate risk stress test that was conducted in 104 significant European banks mentions that: "*Overall, banks have started to integrate climate risk into their stress-testing frameworks. Nevertheless, the majority of supervised institutions are still at a very early stage in the development and implementation of such frameworks. Around 60% of banks do not yet have a well-integrated climate risk stress-testing framework, and most of those banks envisage a medium to long-term time frame for incorporating physical and/or transition climate risk into their framework*". In a similar manner, asset and fund managers are being required by regulators and investors to embed sustainable investment throughout their business models and to consider the ESG activities of the companies to which they invest. As a result of the changes in the priorities of financial intermediaries, non-financial firms must demonstrate transparency, and good performance in ESG-related matters, in order to achieve higher ESG ratings and enjoy better access to funding with more favourable terms. Furthermore, companies that understand the implications of sustainability and integrate ESG into their business models can gain a competitive advantage, not only through new products and services but also through lower reputational risk and enhanced customer satisfaction.

The above changes in the way of doing business in the financial industry motivated us to propose the present edited book to Palgrave Macmillan. We would like to thank Professor Philip Molyneux (Series Editor of the Palgrave Macmillan Studies in Banking and Financial Institutions) and Tula Weis (Executive Editor, Scholarly & Professional Finance, Head of Economics and Finance, Palgrave Macmillan) for the encouragement to go ahead with this project and for their continuous support. We would also like to thank other staff members of Palgrave Macmillan involved in the processing of the manuscript. Needless to say, we are grateful and indebted to all the authors that contributed to this edited work.

The outcome is a book that covers various topics, such as the changing role of banks in the financial system, ESG issues as strategic components of long-term success of financial institutions, the association of corporate social responsibility with customer satisfaction and customer loyalty in the case of banking institutions, the incorporation of ESG factors in fund management investment strategies, the relationship between firm ESG practices and the terms of bank lending, the incorporation of ESG criteria in credit ratings, the politics of climate finance and ESG-related national and international policy initiatives, and the increasing role of bank regulators in the promotion of green and climate finance.

The Changing Role of Banks in the Financial System: Social Versus Conventional Banks by **Simon Cornée**, **Anastasia Cozarenco,** and **Ariane Szafarz** discusses the changing role of banks while focusing on the differences between social and conventional banks. In recent years, there has been an increased interest in these so-called social banks that includes "alternative", "ethical", "green", "sustainable", "values based" banks, etc, hence, being a group of banks that give particular emphasis on environmental and social values while trying to be sustainable. These banks currently have a small market share, but this is expected to grow in the future. Cornée, Cozarenco and Szafarz start by listing the principles and values claimed by social banks, and then outline how these principles are put into practice. Finally, they provide an up-to-date discussion of the academic literature in the field.

ESG Issues as Strategic Components of Long-term Success of Financial Institutions: Are There Differences in Financial Performance and Firm Value? by **Olaf Weber** outlines the importance of ESG issues as a strategic component of long-term success of financial institutions. Weber starts with a historical overview and then outlines different ESG-related products and services. Afterwards, he discusses how the consideration of ESG criteria in financial activities has not only an environmental and societal effect, but also financial implications like a positive effect on financial risk and on business opportunities, as well as the creation of additional income for banks and investment firms. Yet, as he concludes, while ESG should become strategic components of financial institutions, integrating them into their business strategy is not without challenges.

Corporate Social Responsibility, Customer Satisfaction, and Customer Loyalty in Banking Institutions: A Literature Review by **Stratos Kartsonakis** and **Evangelos Grigoroudis** discusses another reason for which banks should place particular emphasis on corporate social responsibility (CSR). In more detail, Kartsonakis and Grigoroudis review the academic literature on the relationship between CSR, customer satisfaction, and loyalty in banking over the period 2009 to 2021. Overall, the reviewed studies point to the conclusion that CSR initiatives of banking institutions have a positive effect on customer satisfaction and customer loyalty. As mentioned by Kartsonakis and Grigoroudis, it seems that CSR becomes an effective way to enhance satisfaction and loyalty as customers are becoming more aware and pay more attention to societal obligations. Therefore, banks may build strong and long-term relationships with their customers, resulting to higher market value and enhanced organizational performance. Kartsonakis and Grigoroudis conclude by suggesting some avenues for future academic research.

Socially and Environmentally Responsible Investments and Mutual Funds by **Michalis Doumpos, Marianna Eskantar** and **Constantin Zopounidis** focuses on the incorporation of ESG factors in the context of designing and implementing social and environmental fund management investment strategies (SRI). This is a particularly important topic since a 2022 report by PwC projects that asset managers globally are expected to increase their ESG-related assets under management to US$33.9tn by 2026, from US$18.4tn in 2021. Doumpos, Eskantar and Zopounidis first outline the principles of SRI and ESG. The authors then discuss the historical developments in the field and present a list of various ESG criteria. Finally, they provide an overview of the recent literature focusing on studies that examine the characteristics of socially responsible mutual funds and their performance, as well as behavioural and portfolio optimization issues.

Firm ESG Practices and the Terms of Bank Lending by **Mingying Cheng** and **Iftekhar Hasan** focuses on the implications of firm ESG practices for the terms of bank lending. Cheng and Hasan start by discussing the underlying theories as to why banks change their lending terms contingent on borrowers' ESG profiles. Then, they outline the channels through which firms' ESG activities affect bank lending decisions (e.g. loan interest rates, maturity, collateral requirements), and how banks may affect borrowers' ESG policies and investment via lending relationships. Finally, Cheng and Hasan provide new empirical evidence that banks' ESG risks become more value-relevant in the capital market over time.

ESG and Credit Risk by **Chrysovalantis Gaganis**, **Fotios Pasiouras** and **Menelaos Tasiou** discusses the incorporation of ESG factors into credit risk assessments. In the first part of their chapter, the authors discuss academic work in the field that provides empirical evidence of the ESG-credit risk nexus. In the second part, they approach the subject from a practitioner's point of view and discuss the approaches of the three major rating agencies (Fitch, Moody's, S&P) and how they incorporate ESG aspects into their credit ratings.

The Politics of Climate Finance and Policy Initiatives to Promote Sustainable Finance and Address ESG Issues by **Paola D'Orazio** focuses on the politics of climate finance and the national and international policy initiatives that promote sustainable finance and address ESG issues. She provides an interesting timeline of the adoption of international agreements aimed at fostering climate finance and the promotion of ESG factors that dates back to 1997 with the Global Reporting Initiative. She discusses how this arena develops over time, referring to other well-known initiatives like the Paris Climate Agreement, the United Nations Sustainable Development Goals, and the EU Action Plan on Sustainable Finance. Along the way she also discusses academic studies on the impact of these initiatives on financial markets, including the green bond market growth. D'Orazio highlights the lack of data to understand climate risks and growing greenwashing concerns and concludes by emphasizing the need for a standardized and mandatory disclosure and reporting system.

Finally, The Role of Bank Regulators in the Promotion of Green and Climate Finance by **Paola D'Orazio** discusses the role of bank regulators in the promotion of green and climate finance. After outlining the relevance of climate risk for financial stability and monetary policy, she refers to the adoption of climate-related financial policies around the world over the period 2000–2020. She then discusses the challenges and shortcomings of existing prudential frameworks, and the consequences of climate change for monetary policy. D'Orazio concludes by offering interesting suggestions for future research and policy directions.

Rethymno, Greece	Chrysovalantis Gaganis
Montpellier, France	Fotios Pasiouras
Portsmouth, UK	Menelaos Tasiou
Chania, Greece	Constantin Zopounidis

Contents

The Changing Role of Banks in the Financial System: Social Versus Conventional Banks 1
Simon Cornée, Anastasia Cozarenco, and Ariane Szafarz

ESG Issues as Strategic Components of Long-term Success of Financial Institutions: Are There Differences in Financial Performance and Firm Value? 27
Olaf Weber

Corporate Social Responsibility, Customer Satisfaction, and Customer Loyalty in Banking Institutions: A Literature Review 47
Stratos Kartsonakis and Evangelos Grigoroudis

Socially and Environmentally Responsible Investments and Mutual Funds 71
Michalis Doumpos, Marianna Eskantar, and Constantin Zopounidis

Firm ESG Practices and the Terms of Bank Lending 91
Mingying Cheng and Iftekhar Hasan

ESG and Credit Risk 125
Chrysovalantis Gaganis, Fotios Pasiouras, and Menelaos Tasiou

The Politics of Climate Finance and Policy Initiatives to Promote Sustainable Finance and Address ESG Issues 145
Paola D'Orazio

The Role of Bank Regulators in the Promotion of Green and Climate Finance 173
Paola D'Orazio

Index 197

Notes on Contributors

Mingying Cheng is a fifth-year Ph.D. student at Fordham University. Her research interests are primarily in the areas of Environmental, Social, and Governance (ESG) and Capital Markets. Her papers have been presented at several reputable academic conferences. She holds a B.B.A. in Finance and an M.S. in Quantitative Economics.

Simon Cornée is an associate professor at the Faculty of Economics and a researcher at CREM CNRS at the University of Rennes 1 (France). He is also associated to the Centre for European Research in Microfinance (CERMi). He specialises in social finance and cooperative banking. He is also interested in the governance and impact of cooperatives and social enterprises. He has published book chapters and articles in, e.g., *Journal of Business Ethics*, *Journal of Money, Credit and Banking*, *Journal of Small Business Management*, *Journal of Theoretical and Institutional Economics*, and *Kyklos*. He received in 2016 the Warren Samuels Prize awarded by the Association for Social Economics at the ASSA Meetings.

Anastasia Cozarenco is an associate professor of economics at Montpellier Business School. She specializes in social finance and microfinance with special attention to regulation, subsidization, discrimination, and gender. She holds the Social and Sustainable Finance Chair funded in partnership with Caisse d'Epargne Languedoc Roussillon and BNP Paribas. She is the Director of the Yunus Center for Social Business and Financial Inclusion. She has published articles in, e.g., *Journal of Business*

Ethics, *Journal of Banking and Finance*, *Applied Economics*, *Small Business Economics*, and *Economics Letters*, and received in 2019 the Warren Samuels Prize awarded by the Association for Social Economics at the ASSA Meetings.

Paola D'Orazio is a researcher at the Chair of Macroeconomics at the Ruhr-Universität Bochum (Germany). Her research interests primarily focus on central banks and financial regulators' role in taming financial instability and promoting a low-carbon transition and the nexus among finance, environmental innovation, and green growth from a complexity economics perspective. She serves as an associate editor of the *Eurasian Economic Review* and *PLOS Sustainability and Transformation*.

Michalis Doumpos is a full professor of Operations Research at the Technical University of Crete, Greece. His research interests include multicriteria analysis, decision support systems, machine learning, and financial risk management. He has published over 100 research articles in premier journals, and he has co-authored several books, edited volumes, book chapters, and conference papers. He is the recipient of the Gold Medal Award of the International Society on Multiple Criteria Decision Making (2019) and he has extensive consulting experience with the financial industry on projects related to credit scoring systems, banking risk management, and real estate appraisal.

Marianna Eskantar is a Ph.D. student at the School of Production Engineering and Management of the Technical University of Crete, Greece. He holds a bachelor's degree in applied mathematics from the University of Crete and an M.B.A. from the Technical University of Crete. Her research interests involve ESG investments, corporate social responsibility, financial management, and multicriteria analysis.

Chrysovalantis Gaganis is a professor of Finance at the Department of Economics at the University of Crete, and a board member of the Financial Engineering and Banking Society. His research interests are in the fields of banking and insurance regulations, corporate governance, credit ratings, and auditing decisions. He has published nine books and over 50 papers in various journals such as: *Management Science*; *British Journal of Management; Journal of Banking & Finance; Journal of Financial Stability*. He has guest edited special issues for various journals and he is an Associate Editor of the *European Journal of Finance*.

Evangelos Grigoroudis is a professor in the School of Production Engineering and Management of the Technical University of Crete, Greece. He currently serves as Dean of the School of Production Engineering and Management of the Technical University of Crete, and he is the national representative in EURO (Association of European Operational Research Societies) and IFORS (International Federation of Operational Research Societies). He is the vice-president of the Hellenic Operational Research Society (HELORS) and member of the American Society for Quality (senior member) and the Sigma Xi Scientific. His research interests include quality management processes, customer satisfaction measurement, multicriteria decision analysis, innovation management, and sustainability assessment.

Iftekhar Hasan is a university professor and serves as the E. Gerald Corrigan Chair in Finance at the Gabelli School of Business at Fordham University. Professor Hasan is a committed interdisciplinary researcher with a significant publication record in top-tier academic journals in Finance, Economics, Accounting, Management, Operation Research, and Information Technology. See the Gabelli School of Fordham University websites, Google Scholar, Ideas/Repec, and Research.com for details on his activities, publications, and impact on the scientific profession.

Stratos Kartsonakis is a Ph.D. student at the Technical University of Greece. He holds an M.Sc. in Technology and Operations Management from the University of Groningen, while he obtained his diploma degree in Production Engineering and Management from the Technical University of Crete. During his studies, his has co-authored research papers in scientific journals and books, and he has presented his work in several international conferences and symposiums. His research interests include multicriteria decision analysis, operational research, decision support systems, customer and employee evaluation and data analysis.

Fotios Pasiouras is Professor of Banking & Finance at Montpellier Business School, and Secretary of the BoD of the Financial Engineering & Banking Society. His research focuses on risk, performance, corporate governance and regulations in the financial industry. It has appeared in around 90 papers published in international journals including the: *Management Science*; *Journal of Money, Credit & Banking*; *Journal of Banking & Finance*; *Journal of Business Ethics*; *British Journal of*

Management. He has received various awards and he is being consistently included in the Stanford's list of the World's Top 2% most-cited scientists by John Ioannidis et al.

Ariane Szafarz is a professor of finance at the Université Libre de Bruxelles (ULB), SBS-EM & CEBRIG, and a co-director of the Centre for European Research in Microfinance (CERMi). Dr. Szafarz specializes in social finance and social banking. She has published books, chapters, notes and articles in, e.g., *Academy of Management Review*, *Econometric Theory*, *European Economic Review*, *Journal of Banking and Finance*, *Journal of Business Ethics*, and *Review of Finance*, and received two times (in 2016 and 2019) the Warren Samuels Prize awarded by the Association for Social Economics at the ASSA Meetings.

Menelaos Tasiou is a senior lecturer in the School of Economics, Accounting and Finance at the University of Portsmouth (UK) and a committee member of its Centre for Sustainable and Innovative Finance. His research lies in the intersection of Finance, Decision Analysis and Management Science and revolves around quantitative modelling of decision-making and empirical corporate finance and banking applications. His publications appear in journals such as, among others, the *British Journal of Management*, *European Journal of Operational Research*, *Journal of Business Ethics*, and *OMEGA*.

Olaf Weber is a professor at the School of Environment, Enterprise and Development (SEED), University of Waterloo, Canada, and holds a University Research Chair in Sustainable Finance. His research and teaching address different fields of sustainable finance, such as sustainable lending and credit risk management, climate finance, sustainable banking, and sustainable corporate finance. In his research, he focuses on the impact of finance and the financial industry on sustainable development.

Constantin Zopounidis is Full Professor of Financial Engineering and Operations Research at the Technical University of Crete, Greece, Distinguished Research Professor at Audencia Business School, France, and Senior Academician of the Royal Academy of Doctors and the Royal Academy of Economics and Financial Sciences of Spain. He has published more than 100 books/edited volumes and more than 300 research papers in scientific journals, edited volumes, conference proceedings and encyclopedias. Prof. Zopounidis was the recipient of the Edgeworth-Pareto Award from the International Society of Multicriteria Decision Making, in 2013. He serves as the Editor-in-Chief and member of the editorial board of several research journals.

List of Figures

The Changing Role of Banks in the Financial System: Social Versus Conventional Banks

Fig. 1 Market share (%) of social banks in Europe 8
Fig. 2 Social banking 17

Firm ESG Practices and the Terms of Bank Lending

Fig. 1 Average number of ESG incidents covered by media 105
Fig. 2 Banks' mean environmental, social, and governance risk exposures over time 105
Fig. 3 Value relevance of earnings and the book values of equity over time 106
Fig. 4 Moderating effect of ESG risks on EPS 110

ESG and Credit Risk

Fig. 1 Interest evolution in the term 'ESG' in academic studies and press. *Source* Authors' elaboration on data from SciencedDirect.com (published articles) and Google Trends (News searches). Illustrated data processed through the 'max' normalization technique for comparability 127

The Politics of Climate Finance and Policy Initiatives to Promote Sustainable Finance and Address ESG Issues

Fig. 1 Timeline of the adoption of international agreements aimed at fostering climate finance and promoting environmental, social, and governance factors (*Source* Author elaboration) 151

Fig. 2 Total adoption of green financial guidelines (GFG), Disclosure Requirement for Financial Institutions (FI) and Non-financial Institutions (NFI) in G20 countries in the period 2000 to 2021 (*Source* Author elaboration) 160

Fig. 3 Cumulated number of green financial guidelines (GFG), Disclosure Requirement for Financial Institutions (FI), and Non-financial Institutions (NFI) in G20 countries in the period 2000 to 2021 (*Source* Author elaboration) 161

The Role of Bank Regulators in the Promotion of Green and Climate Finance

Fig. 1 Overview of the five policy areas considered in the analysis (*Source* D'Orazio and Thole [2022]) 178

Fig. 2 Total number of policies adopted yearly at the global level; time span: 2000–2020 (*Notes* GPP: Green Prudential Policy; OGD: Other Green Disclosure Req.; GFG: Green Financial Guidelines; GB: Green Bonds; GCA: Green Capital Allocation. *Source* D'Orazio and Thole [2022]) 179

Fig. 3 Climate-related financial policymaking in G20 countries: cluster structure, number of policies, and share of policy adoption by type (*Source* D'Orazio [2022]) 180

Fig. 4 Overview of the different waves of adoption (*Source* D'Orazio [2022]) 181

List of Tables

The Changing Role of Banks in the Financial System: Social Versus Conventional Banks

Table 1 Market shares (%) of social banks in 1998 and 2012 in European countries 7

Corporate Social Responsibility, Customer Satisfaction, and Customer Loyalty in Banking Institutions: A Literature Review

Table 1 Dimensional structure of CSR 51
Table 2 Dimensions of CSR in the literature 61
Table 3 Summary of the main relationships tested in the literature 63

Firm ESG Practices and the Terms of Bank Lending

Table 1 Summary statistics 107
Table 2 OLS Regressions of *Stock Price* on *EPS*, *BVPS*, and *ESG Risk Exposure* and Controls 109
Table 3 Evolution in banks' ESG ratings and lending relationship 113
Table A1 The value-relevance of bank ESG risks 116
Table A2 Banks ESG and lending relationships 117

The Changing Role of Banks in the Financial System: Social Versus Conventional Banks

Simon Cornée, Anastasia Cozarenco, and Ariane Szafarz

1 Introduction

The financial sector landscape has seen the development of a new mindset on banking in line with social, environmental, and ethical values. This chapter puts together the diverse financial institutions created along these views. For simplicity, we group them under the label of "social banks (SBs)", while acknowledging the significant heterogeneity of social

S. Cornée
CREM CNRS, and Center for European Research in Microfinance (CERMi), Université de Rennes 1, Rennes, France
e-mail: simon.cornee@univ-rennes1.fr

A. Cozarenco (✉)
Montpellier Business School and CERMi, Montpellier, France
e-mail: a.cozarenco@montpellier-bs.com

A. Szafarz
SBS-EM, CEBRIG and CERMi, Université Libre de Bruxelles (ULB), Bruxelles, Belgium
e-mail: Ariane.szafarz@ulb.be

C. Gaganis et al. (eds.), *Sustainable Finance and ESG*, Palgrave Macmillan Studies in Banking and Financial Institutions,
https://doi.org/10.1007/978-3-031-24283-0_1

missions that can be pursued by these institutions. SBs are at the intersection of two sets: the large set of banking institutions on the one hand, and the value-based grassroots initiatives aimed at addressing financial operations by prioritizing non-financial outcomes on the other hand (Mersland et al., 2019). The latter category includes various entities, such as crowdfunding platforms, microfinance institutions, financial cooperatives, community development banks, and charitable foundations (Cornée et al., 2022), which are concerned with providing funding opportunities and "put the person at the center of the intervention" (Milano, 2011). In short, SBs are financial intermediaries paying attention to the consequences of their financial operations on society and nature (Benedikter, 2011; Cornée & Szafarz, 2014; Paulet et al., 2015; Weber & Remer, 2011). SBs are therefore hybrid organizations (Billis, 2010) in the sense of contributing to the common good while seeking to be financially sustainable.

This chapter focuses on SBs as financial institutions. SBs differ from social impact investing, which is, according to Rizzi et al. (2018), another "dominant" form of social finance. While both social banking and social impact investing pursue social goals, impact investing has specific strategies about fund allocation and client services, seeking to combine social outcomes and financial returns, whereas SBs introduce social values into their business activities; their ethical principles lead to financing social initiatives and generate fair financial returns (Rizzi et al., 2018). Only a few socially responsible investment funds meet the holistic ethical needs of SBs (Krause & Battenfeld, 2019). Impact investing is considered in another chapter of this book.

2 Social Banks in the Field

The roots of social banking can be traced back to the late middle age. According to Milano (2011), the Monti di Pietà,[1] started in the fifteenth century in Italy and spread later to the rest of Europe, can be considered as first examples of SBs. The *raison d'être* of Monti di Pietà consisted in combatting usury practices and providing low-income individual with financial services at fair prices. In the second half of the eighteenth

[1] Monte means a combination of loans, while Pietà refers to an image of Passion of Christ (Milano, 2011).

century, the first savings banks appeared in Europe and secured saving schemes to the middle and low classes, without any speculative intention.

Lending money to the poor is currently considered as the mantra of microfinance institutions, which are mainly active in the Global South, but still exist also as a niche market in rich countries. Most microfinance institutions are subsidized (D'Espallier et al., 2013). A minority of microfinance institutions have the legal status of banks and are called microbanks. Microbanks fall under the definition of SBs. Yet, many microfinance institutions have other statuses, such as NGO (nongovernmental organizations) and NBFI (non-bank financial institutions), which strictly speaking leave them aside from the banking sector (Périlleux et al., 2012; Tchakoute-Tchuigoua, 2010) even though transitioning to banks is a possible evolution (D'Espallier et al., 2017).

In the middle of the nineteenth century, the first cooperative models emerged in England with the Rochdale society (1840) and in France with Philippe Bouchez (1830–1840) and Louis Blanc (1848). At the same time, the cooperative banking movement gained momentum in Germany. This evolution was triggered by the industrial revolution that weakened small business holders and craftspersons in urban areas and the disbanding of the ancient feudal system that plunged peasants and rural residents into misery. During this turmoil period, two important German figures, Hermann Schulze-Delitzsch and Friedrich Raiffeisen, laid the foundations of modern cooperative credit (Cornée et al., 2018). The two men were motivated by distinct ideologies: Raiffeisen was inspired by Christian values, while Schulze-Delitzsch, a liberal, perceived cooperation as a means of offering equal opportunities rather than equality per se. Regardless, the common objective of the nascent cooperative banks was defying usury and offering fair lending opportunities to low-income groups (Guinnane, 1997, 2001, 2003).

This successful cooperative movement spread to the whole world. Today, many SBs are governed as cooperatives, but the cooperative status alone does not guarantee that a bank is social. The divide between social and cooperative banks is subtle. On the one hand, current cooperative banks are primarily oriented toward their members' interest (*mutual interest*) by facilitating credit availability and forging long-term clientele relationships (Périlleux & Nyssens, 2017). They do not take advantage of their bargaining power to "hold up" their borrowers (Angelini et al., 1998). On the other hand, social banks fund projects of general interest with attractive conditions and promote the common good (Cornée &

Szafarz, 2014). Beyond the motivation of helping poor people escape predatory lending, ethics in banking is observable through everyday fair practices toward customers and other stakeholders, transparency of operations, and refraining from excessive speculation (Cornée et al., 2016; Kalmi, 2014).

Based on Gui's (1991) classical distinction between *mutual interest* and *general interest* in the third sector, one could therefore argue that, even though SBs have prolonged the historical missions of cooperative and savings banks and made more explicit their social missions, today's cooperative banks do not automatically qualify as SBs. This distinction does not preclude cooperative banks in their pursuance of mutual interest from generating positive externalities by stabilizing the financial sector (Hesse & Čihák, 2007), smoothing monetary policy contractions (Ferri et al., 2014), and contributing to reducing inequalities (Brei et al., 2018; Minetti et al., 2021).

Despite their predominantly European roots, SBs have later developed in most regions of the world. The first US social bank, the *Shorebank* (formerly *South Shore Bank*), a community development bank, was founded in 1973 in Chicago, followed by the Wainwright Bank and Trust Cy in the 1980s (Benedikter, 2011). Yet, US and European SBs are quite different. According to Benedikter (2011), there are two main differences between US and European SBs which relate to the founding impulse and the meaning of being social. First, SBs in the United States emerged mainly as local initiatives, while their European counterparts tend to have a general scope addressing broad societal issues. Second, US prosocial institutions are typically associated with charities helping disadvantaged people. In Europe, "social" encompasses issues related to environment, technology, and culture, which concern large sections of the population.

Most SBs belong to at least one network identified as a professional association of SBs: FEBEA (*Fédération Européenne des Banques Ethiques et Alternatives*), INAISE (*International Association of Investors in the Social Economy*), and GABV (*Global Alliance for Banking on Values*). FEBEA[2] is a European non-profit organization, created in 2001 to develop and promote ethical finance principles. It was founded by six SBs: *Crédit Coopératif* (France), *Caisse Solidaire du Nord Pas-de-Calais* (France), *Crédal* (Belgium), *Hefboom* (Belgium), *Banca Etica* (Italy), and *TISE*

[2] https://febea.org/.

(Poland). Nowadays, it gathers 33 financial institutions from 15 European countries. According to FEBEA's website, the core values of social banking rely on five basic principles: (1) Money serves the common good; (2) Transparent use of money for the real economy; (3) Credit add value by supporting social economy and social entrepreneurship; (4) Avoid speculation and reinvest profits in line with social objectives. These principles imply that SBs should focus on funding cultural, social, and environmental projects.

INAISE[3] is a global network of socially and environmentally oriented financial institutions. Created in 1989, INAISE grew rapidly as social finance gained importance in visibility and volume of activity worldwide. It has currently 23 members active in 19 countries. INAISE promotes transparency, trust, equal and fair access to finance, quality of the services, sustainability, cooperation, democracy, local footprint. INAISE focuses on social investing, but only four of its members have a bank legal status (*BMS S.A.* in Mali, *Caisse d'Economie Solidaire Desjardins* in Canada, *Ecology Building Society* in the UK, and *Triodos Bank* in the Netherlands).

GABV[4] was founded in 2009 by ten banks inspired by a common aspiration for a fairer financial system. Today, the organization is present in 40 countries. According to GABV (2020), 29% of its 63 members are based in Europe, 22% are active in North America, 22% are in Asia and the Pacific, 21% are in Latin America, and the remaining 6% are in Africa. The six guiding principles of GABV membership are: social, environmental, and sustainable impact; financing real economy; long-term relationships with clients; self-sustainability and resilience; transparent and inclusive governance; and embeddedness of these values in the culture of the bank.

Despite their historical background, SBs attracted public interest only since the 2007 financial crisis. The massive involvement of banks in suspicious operations on mortgage-backed securities and other obscure derivative products lead the public to realize there was a critical need to align the management of financial institutions with ethical principles. The crisis also showed that, in contrast to several conventional banks, which faced high losses and bankruptcy, SBs were insulated from the detrimental consequences of the crisis. In fact, the assets of European SBs increased by 20–25% per year on average during the 2006–2008 period

[3] http://inaise.org/en/.

[4] https://www.gabv.org/.

(Benedikter, 2011). Between 2007 and 2012, the net income of SBs experienced an average 16% annual growth rate (Weber, 2013) and their asset quality improved significantly compared to large conventional banks (Mykhayliv & Zauner, 2018). This remarkably resilient growth during the crisis can be explained by clients of conventional banks realizing that the highly speculative operations of their banks were putting their own savings at risk. As a consequence, these savers shifted their assets toward SBs, which were accurately perceived as safer. Valls Martínez et al. (2020) observe a similar trend during the period that stretched from 2015 to 2018.

To illustrate how instrumental the financial crisis was to the economic development of social banking, we build on the set of European SBs identified by Cornée et al. (2020). Table 1 and Fig. 1 provide complementary perspectives on the impressive growth of SB market shares in Europe, the stronghold of social banking between 1998 and 2012. Table 1 compares national market shares of SBs in 1998 and 2012. Figure 1 shows the global evolution of yearly market shares averaged across countries. The figure shows that SBs experienced a remarkable growth pace over the 1998–2012 period, as their market share was multiplied by four (from 0.10 to 0.40%). Despite their growing popularity, SBs are still small players in the financial landscape.

Recent professional accounts corroborate the successful evolution of SBs beyond the borders of Europe. GABV reports that, over the 2010–2019 period, the average SB annual asset growth rate was 15.2%, to be compared to 2.7% for global systemic banks (GABV, 2020). Recent work suggests that the upwards trend is there to stay. For Germany, Krause and Battenfeld (2019) view the potential market size of SBs between 10 and 26% of adult population, while Mykhayliv and Zauner's (2018) estimation of 15.2 million potential customers sits closer to the lower bound of the interval.

3 Empirical Studies on Social Banks

Identifying SBs is a challenging task (Cornée et al., 2020; Karl, 2015; San-Jose et al., 2011). By design, SBs differ from conventional banks by promoting social objectives in lending, but there is also significant heterogeneity among SBs. Social banks are identifiable from conventional banks in several dimensions, among which: a specific target clientele,

Table 1 Market shares (%) of social banks in 1998 and 2012 in European countries

Country	*Social banks*	*Share in total banking assets (%)*	
		1998 (%)	*2012 (%)*
Denmark	Andelskassen OIKOS, Folkesparekassen*, Merkur—Den Almennyttige Andelskasse*	0.015	0.051
France	Crédit Coopératif*, La Nef*	0.203	0.224
Germany	Bank für Sozialwirtschaft Aktiengesellschaft*, GLS Gemeinschaftsbank eG*, IntegraBank eG München*, Ökobank eG*, Steyler Bank GmbH, UmweltBank AG*	0.078	0.192
Italy	Banca Popolare Etica SPA*, Cassa Padana Banca di Credito*, Cassa Rurale di Bolzano Soc. Cooperativa*, Eticredito-Banca Etica Adriatica SpA	0.000	0.111
Malta	APS Bank Limited*	n/a	n/a
Netherlands	Algemene Spaarbank voor Nederland—ASN Bank NV, Triodos Bank NV*	0.591	0.885
Norway	Cultura Sparebank*	0.000	0.013
Spain	Caja Laboral Popular Coop. de Credito, Colonya, Caixa d'Estalvis de Pollença*	0.000	1.175
Sweden	Ekobanken medlemsbank*	0.000	0.010
Switzerland	Alternative Bank Schweiz ABS*, Freie Gemeinschaftsbank BCL*	0.023	0.062
UK	CAF Bank Ltd, Charity Bank Limited (The)*, Co-operative Bank Plc (The), Ecology Building Society (The)*, Reliance Bank Limited	0.062	0.902
Mean		0.097	0.363

alternative risk management techniques adapted to addressing the challenges raised by the special asymmetric information stemming from this clientele, transparent information and simple intermediation, and stakeholder participation in decision-making. The remaining of this section will address these dimensions through the lens of empirical evidence. The purpose is to assess the practical consequences for banking activity of committing to social values.

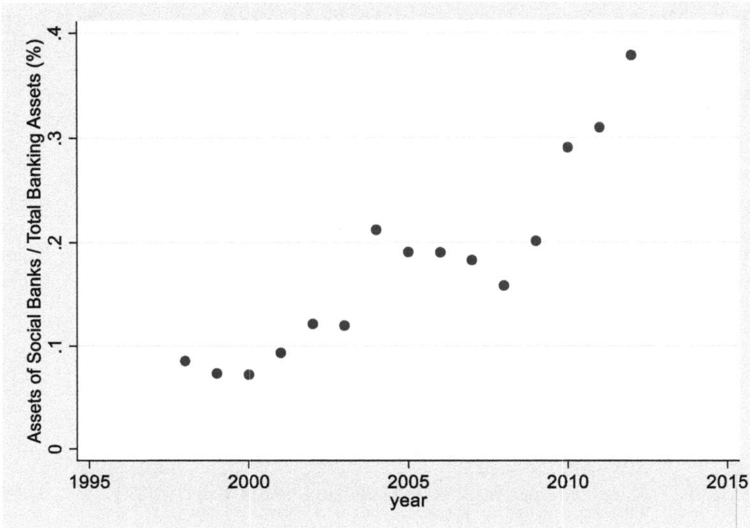

Fig. 1 Market share (%) of social banks in Europe

Target Clientele

Asset side: Borrowers. The financial transactions of SBs focus on funding the real economy rather than trading in speculative markets (Cornée et al., 2016). SBs provide services to social enterprises (Defourny, 2014), non-profit organizations, and community-oriented projects to boost local development (Périlleux, 2015). These endeavors are evidently less profitable than their for-profit counterparts. Although social enterprises sometimes make profits, their main goal is not profit maximization (Besley & Ghatak, 2017). Social enterprises typically promote financial inclusion of disadvantaged people, women's empowerment, fair trade, clean energies and recycling, community services, and so on (Di Domenico et al., 2010). Yet, they can vary substantially (Borzaga & Defourny, 2001). According to Defourny and Nyssens (2008, 2010), social enterprises are defined by the combination of entrepreneurial nature and social orientation. Due to these characteristics, social enterprises find it difficult to attract funding from conventional banks, and therefore seek preferably financing from prosocial funders, such as social banks.

Conventional banks are ill-adapted to funding prosocial entities, which are typically both less remunerative and more informationally opaque than projects undertaken by for-profit firms, thereby leading to severe credit rationing for projects undertaken by social economy actors. This hypothetical mismatch offers a rationale for SBs. By studying Banca Etica, a large SB operating in Italy, with more than half of its borrowers being not-for-profit entities, Becchetti et al. (2011) provide evidence corroborating the hypothesis. They show that roughly 20% of the loans granted by Banca Etica are subject to rationing (i.e., the amount disbursed is lower than the amount requested by the borrower), while loans are denied to 15% of the applicants. This denial rate is low compared to that of conventional banks (Minetti & Zhu, 2011), confirming that SBs facilitate access to credit for borrowers otherwise redlined or rationed.

SBs also seek to alleviate indirect forms of credit rationing. In this respect, low collateralization is claimed to be a distinctive feature of social banking. While mainstream banks generally require collateral from the vast majority of their small-business borrowers, Becchetti and Garcia (2011) report that *Banca Etica* has 42% of uncollateralized loans. The requested collateralization depends positively on *ex ante* risk, and negatively on the existing relationship with the borrowers. The authors also show that the SB counterbalances low collateral requirements by maintaining close connections with umbrella organizations of social enterprises. Likewise, most microfinance institutions require no formal collateral from their borrowers. Instead, they rely on social collateralization, a mechanism particularly relevant for group lending (Postelnicu et al., 2014).

Liability side: Depositors and Users of other Banking Services. Krause and Battenfeld (2019) provide novel evidence on the pool of potential users of the services proposed by social banks. Based on a survey with German respondents who are customers of either SBs or conventional banks, their results show that the two groups differ significantly in gender, age, and education. Male customers are more likely associated with SBs than female ones. The authors suggest that this could have to do with the fact that social banking is in an early stage of development. Moreover, SB customers tend to be younger, more urban, with higher education, but there is no significant difference in income levels. The authors argue that the socioeconomic profile of customers is linked to greater awareness and understanding of social and ecological issues. As opposed to clients of conventional banks, the clients of SBs are also logically more motivated by social returns than by financial returns. This is in line with Bauer and

Smeets's (2015) study documenting that investors with stronger social identification toward their bank—typically highly educated, young, and low-wealth individuals—allocate substantially more of their wealth to this bank. These strong-identifying investors expect low returns from their investment.

Matching the Two Sides: Overcoming Information Asymmetry

Akin to conventional banks, the credit activity of SBs can be confronted to severe information asymmetries (Bhattacharya & Thakor, 1993; Diamond, 1984). Typical devices for addressing these obstacles include screening, selection, and monitoring mechanisms. The social mission of SBs entails two distinct informational issues: (1) assessing the creditworthiness of loan applicants that have relatively opaque and informal functioning based on unconventional economic objectives, and (2) gauging the social commitment and feasibility of the projects to be funded.

Assessing creditworthiness. SBs deal with a specific pool of borrowers whose creditworthiness is uneasy to assess. Like small and medium enterprises (SMEs), which are known to be opaque (Berger & Udell, 2002), social enterprises are difficult to assess with standard, quantitative lending technologies (Farber & Reichert, 2021). The issue is even more acute for social enterprises than for standard SMEs. The financial sustainability of prosocial entities depends on features that are hard to quantify, such as relational capital, acquisition of nonmarket resources, and social value creation (Cornée, 2014). Two factors may further magnify informational opacity. First, social enterprises are often innovative businesses, preventing lenders from using past experience to reduce the informational gap. Second, social enterprises operate on a small scale and anchor their activities in local communities, thereby adding complexity to disclosing business facts.

To tackle these informational problems, SBs typically resort to relational approaches. Relationship lending (Cornée et al., 2012) is regarded as a powerful technology against information asymmetry, particularly when it comes to adverse selection. Equipped with their relational methods, SBs serve borrowers otherwise excluded or severely rationed by the mainstream banking market due to their informational opacity. This unique know-how of social banks helps to fill the market gap and alleviate credit rationing (Stiglitz & Weiss, 1981).

The lending technology of SBs is built on collecting both quantitative and qualitative information. In conventional credit market, the use of quantitative (hard), financial and standardized information has increased at the expense of qualitative (soft) information (e.g., the skills of the entrepreneur, the company's governance, the quality of the project to be financed, etc.). The banking sector has been fully engaged in the information revolution for several decades by developing increasingly complex information systems (Artis & Cornée, 2016). In contrast, SBs gather soft information on opaque credit applicants at the selection stage. They can appraise the creditworthiness of their loan applicants with great accuracy and make well-informed credit decisions. There is surprisingly little academic interest in the beneficial aspects of soft information in creditworthiness assessment. Based on proprietary data retrieved from a French SB, Cornée (2019) shows however that soft—in addition to hard—information increases accuracy of credit default models. For small social enterprises, soft information tends to be even more valuable than hard information. The lending technology based on relationship lending and soft information models relies on the skills of loan officers and underlines the importance of staff retention (Artis & Cornée, 2016; Doering & Wry, 2022; Godfroid et al., 2022).

Still collecting soft information can be tedious and costly. The cost-benefit analysis of Cornée (2019) weighs the pros and cons of using soft information by opposing the cost reduction gained from improved predictive accuracy of defaults to the increase in labor costs. The results suggest that the former effect dominates the latter, thus indicating that collecting soft information is likely valuable. Interestingly, the outcome is larger for firms in a credit relationship with the bank, which eases the collection of soft information and increases its predictive value.

Social Screening. SBs are accountable to their socially minded funders and must show the pro-social accomplishments of the capital they allocate. Hence, in addition to conducting creditworthiness assessments, SBs are bound to assess the social dimension of projects submitted for funding, and often develop social scores in addition to credit evaluation. In sum, SBs spend resources to screen loan applicants on both a financial basis *and* a social basis.

Social credit scoring is bank-specific, which explains why credit granting by SBs is often considered a black box. To open the box, Cornée and Szafarz (2014) exploit a hand-collected data set on the business loans granted by a French social bank, *La Nef*. Each borrower in their sample was graded on both a social scale and a financial scale. In contrast to the financial rating, the social rating, interpreted as a proxy for the proximity between the applicant's and the bank's social identities, is not determined according to strict rules.

To assess the cost of extra workload devoted to social screening, Cornée et al. (2018) use balance-sheet information from European banks. Their results suggest that the operating costs of SBs are not significantly higher than those of their mainstream counterparts, meaning that the extra costs of dual screening might be offset by a cheaper workforce from intrinsically motivated staff, accepting lower wages in exchange for working in a social enterprise. If so, despite labor-intensive dual screening stemming from their mission, SBs manage to avoid excessive costs. Mykhayliv and Zauner (2018) compare the performance of SBs with that of "big banks" and confirm that SBs have lower operating costs than their non-social counterparts.

Transparency and Simple Intermediation

Informational asymmetries go beyond the relationship between a bank and its clients. They also plague the relationship between the bank and its funders, shareholders, and depositors. In social banking, the double bottom line makes it even more demanding to consolidate trust between stakeholders. Empirical evidence shows that SBs operate more transparently than other banks. They carry out fewer speculative and obscure transactions than their purely for-profit counterparts. They also favor direct intermediation by focusing on simple savings and loan products (Mykhayliv & Zauner, 2018). This strategy translates into a higher share of interest income in the total bank's income combined with a higher deposit-to-asset ratio and a lower loan-to-asset ratio, highlighting excess liquidity (Cornée et al., 2016). The difficulty SBs face in transforming deposits into credits could be due to their stringent selectivity reducing lending opportunities. High selectivity leads to the paradoxical situation in which social screening can undermine direct intermediation and transparency. Still, SBs manage excess liquidity prudently relying on simple financial transactions.

Overall, despite being confronted to losses like other banks, SBs are immune to the impact of toxic assets. San-Jose et al. (2011) confirm that European SBs are more transparent to their stakeholders than other banks, and that they preferably use simple intermediation. Most SBs release exhaustive lists of the businesses and individuals they support and disclose the loan characteristics, such as amount and duration, as well as the aim of the funded projects. The websites and annual reports of SBs inform about their asset management. In sum, the empirical literature confirms that SBs adhere to the values and principles promoted by their associations, namely, transparency, trust, and no speculative operations.

Stakeholder Involvement

SBs tend to adopt specific ownership and governance structures while allowing for some diversity in legal statuses: They can be either stakeholder banks or shareholder banks. The most common status of SB stakeholder ownership is cooperative bank, which naturally limits owners' residual claims (Kalmi, 2014). For SBs, having a cooperative status facilitates aligning the interests of the managers with the bank's social mission (Kitson, 1996). Cooperatives also tend to adopt low-risk investment strategies (Hesse & Čihák, 2007).

However, not all SBs are governed by stakeholder ownership. In Cornée et al.'s (2020) sample, 39% of the SBs have a shareholder-owned status associated with specific limitations. For example, shareholders' voting rights at Alternative Bank Schweiz (ABS, Switzerland) and Triodos Bank (The Netherlands and Belgium) are capped. Each ABS shareholder must remain below the 3% voting right threshold. Triodos Bank's shares are held in trust by an ad hoc foundation, whose board is appointed by depository receipt holders with limited voting rights. Both stakeholder ownership and shareholder ownership structure with self-regulatory arrangements help SBs to fulfill the same objective of restricting profit distribution and curbing the power of dominant capital holders. This limitation of ownership claims is instrumental to obviate breaches in the moral contracts between a SB and its stakeholders. For instance, the goodwill of depositors toward the SB could depreciate if they are suspicious about their donations or sacrifices being pocketed by capital holders. Likewise, employees may refrain from accepting below-market wages if they fear that their benevolent efforts serve capital holders' interests.

4 Theorizing the Business Model of Social Banks

The expression "business model" is frequently employed loosely to designate anything that has to do with the functioning of a sector. In contrast, there is an academic consensus on characterizing the so-called "business model" in terms of value creation (Yip & Bocken, 2018). To do so, Osterwalder and Pigneur (2010) use an extensive framework based on nine building blocks: key partners, key resources, key activities, value propositions, customer relationships, customer segments, channels, revenue streams and cost structure. Accordingly, we structure this section in three parts dealing with: the supply side focusing on the key partners and resources of SBs; the demand side addressing the key activities, value propositions, customer relationships and segments served by SBs, and finally the global perspective linking the supply and the demand sides and addressing the channels, revenue streams and cost structure of SBs.

Supply Side: Socially Minded Funders

Most funders of SBs are convinced that doing good does not come for free. In other words, financing social projects requires waiving at least some capital returns. This fact can be theorized, and subsequently tested, for two separate groups of funders: depositors and owners.

The financial sacrifice of SB depositors can be measured by the "social premium" computed as the spread between the interest rate they receive from the SB and the market interest rate obtained from for a similar savings opportunity. Becchetti and Garcia (2011) estimate that the annual social premium conveyed by *Banca Etica* was around 150 basis-points (or 1.5%) in 2007. A panel analysis conducted by Cornée et al. (2020) on 5,400 European banks over the 1998–2013 period confirms the existence of a significant social premium. The recent macroeconomic situation, characterized by negative market rates, is particularly detrimental to the banks largely funded by deposits (Basten & Mariathasan, 2018; Eggertsson et al., 2019). In this context, two SBs, namely *Alternative Swiss Bank* and *Triodos*, were among pioneers in imposing explicitly negative deposit rates.

Regarding owners, SBs, usually implement ownership structures that strongly restrict residual claimants' rights. This governance design can be understood as a credible commitment device toward depositors who accept to receive below-market interest rates. Since financial intermediation is plagued by asymmetric information, SBs need to convince their depositors that their sacrifice is not just another way to increase owners' profits. Empirical evidence indicates that SBs abide by the principles of reduced ownership rights and limited profit distribution. In this regard, San-Jose et al. (2011, p. 152) report that "*the dimension of obtaining benefit refers to good bank management, because ethical banks do not generally distribute benefits between shareholders and, if at all they do so, the distribution is very limited, and profit is, therefore, only residual*". Using return on assets as a measure of owners' remuneration, Cornée et al. (2020) confirm this statement. Their estimation associates SBs with a sizable 20 basis-points deduction in return on assets with respect to a global average of 0.50%.

The financial sacrifice of prosocial investors and depositors is rationalized by their intrinsic motivation. Experimental evidence on reciprocal behavior consistently shows that a large proportion of individuals exhibit social preferences: They care not only for their self-interest, but also for the well-being of others (Gintis et al., 2004). Reciprocity typically arises when individuals are prone to sacrifice their own resources to encourage positive action or punish negative action (Fehr & Gächter, 2000).

Reciprocal motivations are boosted by social identification, that is by a person's sense of self derived from perceived membership to a social group (Akerlof & Kranton, 2005). In social banking, the social premium is likely to be greater when the investor self-identifies with the ethical values promoted by the bank. Riedl and Smeets (2017) elucidates the behavioral micro-foundations of social investment thanks to a unique data set that links administrative data of conventional and social investors to their behavior in controlled experiments and to their answers in a comprehensive survey. The authors find that the intrinsic social preferences revealed through a trust game conducted in lab[5] are correlated with real-life social investments. Compared to conventional investors, social investors have

[5] The authors use a variant of the trust game coined by Berg et al. (1995). The two players are endowed with EUR 50. The first mover (the sender) decides on an amount between EUR 0 and 50 to send to the second mover. The amount sent is tripled by the experimenter, and the second mover (the receiver) decides how much of the money he/she returns to the sender. Hence, the sender's earnings are EUR 50 minus the amount sent back plus the amount returned by the receiver. The receiver's earnings are EUR 50 plus the triple of the amount received from the sender minus the money sent back. The authors use the second mover's behavior as a measure of intrinsic social preferences.

also a higher propensity to donate to charities, suggesting that social investment is not a substitute for charity donations.

In sum, the social contribution of SBs to the common good is unambiguously conditional on investors making financial sacrifices, enabling access to capital at lower cost.

Demand Side: Prosocial Borrowers

Thanks to their access to cheaper funding capital, SBs channel the financial sacrifices of their funders to the lending side of the balance sheet by offering preferential credit conditions to their borrowers. Cornée et al. (2020) show that the average interest rates charged to borrowers in social banks is significantly lower than in conventional banks for same-risk loans. At the micro level, Cornée and Szafarz (2014) find that, all else equal, borrowers with a higher social rating receive loans with lower interest rates and have a lower probability of default. Becchetti et al. (2011) highlight that the fraction of nonperforming loans of SBs is low when compared to that of the conventional banking sector.

Like for funders, reciprocity and social identification help to rationalize the lender-borrower interactions in social banking (Périlleux, 2015). In this case, reciprocity can be theorized as either unconditional or conditional (Cornée et al., 2022). In the theoretical setting proposed by Barigozzi and Tedeschi (2015, 2019), reciprocity is unconditional and simply derives from the nature of SBs. Under unconditional reciprocity, prosocial borrowers perceive an added stream of utility for being granted a loan by a SB, and spontaneously reduce their probability of (strategic) default. In contrast, under conditional reciprocity, the borrowers need a (costly) signal sent by the lender to experience reciprocity. In both cases, the reciprocal borrower will exert more effort to repay the loan (Fehr and Zehnder, 2005; Brown & Zehnder, 2007).

Cornée et al. (2012) provide experimental evidence showing that social bankers charge fairer rates than commercial bankers, thereby reducing the borrowers' propensity to shirk by choosing risky investment projects. Value-sharing between lenders and borrowers is deemed to further strengthen conditional reciprocity. The asymmetric-information model developed by Cornée and Szafarz (2014) encapsulates this view. The SB performs a costly screening to detect the extent to which credit applicants share its social values and adjust its interest rate accordingly. In return, motivated borrowers who realize they benefit from a rebate in interest rate

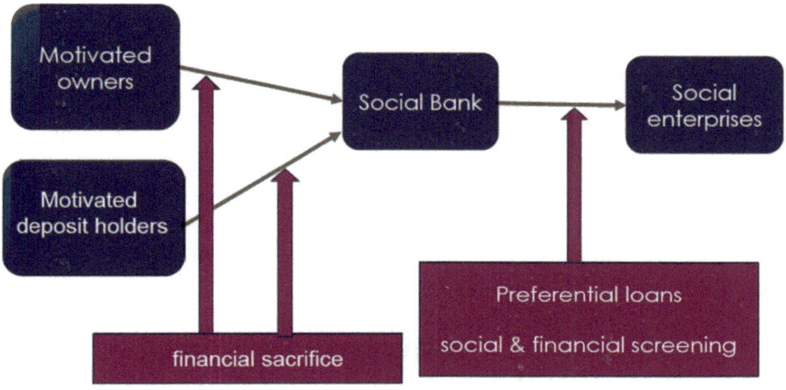

Fig. 2 Social banking

granted by the SB (because of their values) reciprocate, thereby reducing moral hazard and credit defaults.[6]

A Global Perspective on Social Banking

Ultimately, SBs appear as financial institutions bridging the gap between social funders and motivated borrowers. By matching the two sides of impact-based financial intermediation, they promote social values and serve the common good (Cornée et al., 2022). Figure 2 sketches the global picture in which SBs are funded by motivated owners and deposit holders who accept to make a financial sacrifice and provide preferential loans at low interest rates to social enterprises. In brief, they pass their funders' financial sacrifice on to carefully screened borrowers.

Most social SBs charge near-zero interest rates while the least social SBs charge just-below-market rates. Consequently, the leeway of SBs may be represented by an interest rate segment that is circumscribed by a lower limit of zero and an upper limit of the market interest rate. They use social and financial screenings to select their borrowers. The business model

[6] Evidence indicates that any non-financial factor increasing borrower creditworthiness is favorable to risk management (Weber et al., 2010). Chava (2014) also suggests that firms generating less negative environmental externalities benefit from cheaper capital.

presented in Fig. 2 is in line with Cornée et al. (2020) suggesting a two-pillar business model of value-based financial intermediation.

5 Conclusion

This chapter started by listing the principles and values claimed by SBs. Next, it moved to checking how these principles are put into practice. Last, it discussed how social banking is theorized in the academic literature. Overall, there is good news: SBs do walk the talk. The claimed missions of social banking materialize as actual deeds. SBs finance the real economy, they assess the ethicality of loan applicants, they are transparent on their activities, and they refrain from complex, potentially speculative financial operations.

Our literature survey rationalizes the role played by SBs in the global financial landscape with an integrated representation of social banking as value-based financial intermediation addressing a gap in the credit market. Investors that fund SBs, be they capital-holders or depositors, assent to trade values for financial remuneration. The social premium they grant is fundamental for SBs to accomplish their very mission, which consists in carefully selecting social projects and providing them with capital at below-market interest rates. If SBs did not exist, many social projects would remain denied or severely credit rationed.

Being a unique niche, social banking offers an attractive alternative to the big-bank model harshly criticized during the last financial crisis. The remarkable achievement of SBs provides additional proof that grassroots economic initiatives are welcome and can be sustainable, even without financial support from public authorities. From a historical perspective, social banking corresponds to the new generation of financial actors seeking to strike a balance between community/social goals and financial constraints (i.e., breaking even or achieving a modest surplus) in the footsteps of Monti di Pietà, savings banks, and credit cooperatives.

At the institutional level, however, these flourishing alternative banks could benefit from an adapted regulatory framework (Serres, 2019). The current framework poorly fits social banking mainly because the Basel accords force banks to use lending technologies based on hard information, which are at odds with the relational soft-information approach central in financing social economy enterprises (Ferri & Neuberger, 2015; Rajan et al., 2010). In addition, the transition from the originate-to-hold model, in which banks keep the loans they grant in their own balance sheet, to the originate-to-distribute model, whereby loans are sold in structured financial markets, contributes to releasing complex products, such as securitized loans and other financial derivatives (Diamond & Rajan, 2009).[7] Developing such products collide with the core principles of simple intermediation and transparency advocated by SBs.

Some segments of social banking, such as microcredit activities benefit however from a specific regulatory framework that recognizes their prosocial specifies. Regulators in most industrialized countries carefully monitor the activities of microcredit institutions, probably because these institutions receive significant subsidies from national and supranational public authorities and their activities belong to social finance (Cozarenco et al., 2022; Morduch & Ogden, 2019). Key regulatory rules impose ceilings on interest rates and loan sizes (see Cozarenco & Szafarz, 2019, for a detailed lists of obligatory and recommended ceilings in place in North America and Europe). While interest rate caps have long been in use as protection against loan sharks (Caballero-Montes et al., 2021), loan ceilings are less common and they can have unexpected perverse effects, such as encouraging small enterprises that can afford it to combine loans from conventional banks and microcredit institutions, while leaving others with harsher access to credit (Cozarenco & Szafarz, 2020).

In terms of market weight, SBs represent the bulk of social finance organizations. Yet, there exists a broad spectrum of actors participating in non-banking social finance. These alternative vehicles of social finance encompass crowdfunding platforms and a myriad of local community-based financial initiatives, which mostly remain below the radar (Gafni et al., 2021). These initiatives can take the form of informal/local savings groups and associations promoting complementary currencies (Meyer & Hudon, 2017). SBs are distinguished by their legal banking structure and the financial and operational constraints attached to it. Meanwhile, this legal framework has also the advantage of allowing SBs operate at much large scale as a full-fledge financial intermediary. But the question as to which is the global financial structure that would best serve social endeavors, and ultimately the common good, is still open.

[7] In the last two decades and mostly in the United States, financial intermediation largely adopted the originate-to-distribute model. Nevertheless, the potentially complex consequences of originate-to-distribute operations in terms of information and incentives are mainly ignored by the current Basel supervisory framework (Ferri & Neuberger, 2015).

References

Akerlof, G. A., & Kranton, R. E. (2005). Identity and the economics of organizations. *Journal of Economic Perspectives, 19*(1), 9–32.

Artis, A., & Cornée, S. (2016). Composition, interprétation et mémorisation du savoir idiosyncrasique dans la banque solidaire. *Systemes d'information Management, 21*(3), 93–131.

Angelini, P., Di Salvo, R., and G. Ferri (1998). Availability and cost of credit for small businesses: Customer relationships and credit cooperatives. *Journal of Banking & Finance, 22*(6–8), 925–954.

Barigozzi, F., & Tedeschi, P. (2015). Credit markets with ethical banks and motivated borrowers. *Review of Finance, 19*(3), 1281–1313.

Barigozzi, F., & Tedeschi, P. (2019). On the credibility of ethical banking. *Journal of Economic Behavior & Organization, 166*, 381–402.

Basten, C., & Mariathasan, M. (2018). How banks respond to negative interest rates: Evidence from the Swiss exemption threshold.

Bauer, R., & Smeets, P. (2015). Social identification and investment decisions. *Journal of Economic Behavior & Organization, 117*, 121–134.

Becchetti, L., & Garcia, M. M. (2011). Informal collateral and default risk: Do 'grameen-like' banks work in high-income countries? *Applied Financial Economics, 21*(13), 931–947.

Becchetti, L., Garcia, M., & Trovato, G. (2011). Credit rationing and credit view: Empirical evidence from loan data. *Journal of Money, Credit and Banking, 43*(6), 1217–1245.

Benedikter, R. (2011). *Social banking and social finance*. Springer.

Berg, J., Dickaut J., & McCabe, K. (1995). Trust, reciprocity, and social history. *Games and Economic Behavior, 10*, 122–42.

Berger, A. N., & Udell, G. F. (2002). Small business credit availability and relationship lending: The importance of bank organizational structure. *Economic Journal, 112*(477), F32–F53.

Besley, T., & Ghatak, M. (2017). Profit with purpose? A theory of social enterprise. *American Economic Journal: Economic Policy, 9*(3), 19–58.

Bhattacharya, S., & Thakor, A. V. (1993). Contemporary banking theory. *Journal of Financial Intermediation, 3*(1), 2–50.

Billis, D. (Eds.), (2010). Hybrid organizations and the third sector: Challenges for practice, theory and policy. Palgrave Macmillan.

Borzaga, C., & Defourny, J. (Eds.). (2001). *The emergence of social enterprise*. Routledge.

Brei, M., Ferri, G., & Gambacorta, L. (2018). *Finance structure and income inequality* (BIS Working Paper No 76).

Brown, M., & Zehnder, C. (2007). Credit reporting, relationship banking, and loan repayment. *Journal of Money, Credit & Banking, 39*(8), 1883–1918.

Caballero-Montes, T., Godfroid, C., & Labie, M. (2021). Are interest rate caps a relevant tool to cool down overheating microfinance markets? *Strategic Change*, *30*(4), 319–330.

Chava, S. (2014). Environmental externalities and cost of capital. *Management Science, 60*(9), 2223–2247.

Cornée, S. (2014). Soft information and default prediction in cooperative and social banks. *Journal of Entrepreneurial and Organizational Diversity, 3*(1), 89–109.

Cornée, S. (2019). The relevance of soft information for predicting small business credit default: Evidence from a social bank. *Journal of Small Business Management, 57*(3), 699–719.

Cornée, S., Fattobene, L., & Migliorelli M. (2018). An overview of cooperative banking in Europe. In M. Migliorelli (Eds.), *New cooperative banking in Europe. Strategies for adapting the business model post crisis* (pp. 1–27). Springer.

Cornée, S., Jegers, M., & Szafarz, A. (2022). Feasible institutions of social finance: A taxonomy. *Journal of Institutional and Theoretical Economics, 178*(3), 280–310.

Cornée, S., Kalmi, P., & Szafarz, A. (2018). How costly is social screening? Evidence from the banking industry. *Economics Bulletin*, *38*(1), 532–540.

Cornée, S., Kalmi, P., & Szafarz, A. (2016). Selectivity and transparency in social banking: Evidence from Europe. *Journal of Economic Issues, 50*(2), 494–502.

Cornée, S., Kalmi, P., & Szafarz, A. (2020). The business model of social banks. *Kyklos, 73*(2), 196–226.

Cornée, S., Masclet, D., & Thenet, G. (2012). Credit relationships: Evidence from experiments with real bankers. *Journal of Money, Credit and Banking, 44*(5), 957–980.

Cornée, S., & Szafarz, A. (2014). Vive la différence: Social banks and reciprocity in the credit market. *Journal of Business Ethics, 125*(3), 361–380.

Cozarenco, A., Hartarska, V., & Szafarz, A. (2022). Subsidies to microfinance institutions: How do they affect cost efficiency and mission drift? *Applied Economics*, *54*(44), 5099–5132.

Cozarenco, A., & Szafarz, A. (2019). Microfinance in the North: Where do we stand? In M. Hudon, M. Labie, & A. Szafarz (Eds.), *Research Agenda for financial inclusion and microfinance* (pp. 125–137). Edward Elgar Publishing.

Cozarenco, A., & Szafarz, A. (2020). The regulation of prosocial lending: Are loan ceilings effective? *Journal of Banking & Finance, 121*, 105979.

Defourny, J. (2014). From third sector to social enterprise: A European research trajectory. In J. Defourny, L. Hulgård, & V. Pestoff (Eds.), *Social enterprise and the third sector* (pp. 33–57). Routledge.

Defourny, J., & Nyssens, M. (2008). Social enterprise in Europe: Recent trends and developments. *Social Enterprise Journal, 4*(3), 202–228.

Defourny, J., & Nyssens, M. (2010). Conceptions of social enterprise and social entrepreneurship in Europe and the United States: Convergences and divergences. *Journal of Social Entrepreneurship, 1*(1), 32–53.

D'Espallier, B., Goedecke, J., Hudon, M., & Mersland, R. (2017). From NGOs to banks: Does institutional transformation alter the business model of microfinance institutions? *World Development, 89*, 19–33.

D'Espallier, B., Hudon, M., & Szafarz, A. (2013). Unsubsidized microfinance institutions. *Economics Letters*, 120(2), 174–176.

Diamond, D. W. (1984). Financial intermediation and delegated monitoring. *Review of Economic Studies, 51*(3), 393–414.

Diamond, D. W., & Rajan, R. G. (2009). The credit crisis: Conjectures about causes and remedies. *American Economic Review, 99*(2), 606–610.

Di Domenico, M., Haugh, H., & Tracey, P. (2010). Social bricolage: Theorizing social value creation in social enterprises. *Entrepreneurship Theory and Practice, 34*(4), 681–703.

Doering, L., & Wry, T. (2022). The challenges of supporting necessity entrepreneurs: Understanding loan officer exit in microfinance. *Journal of Business Venturing*, *37*(2), 106189.

Eggertsson, G.B., Juelsrud, R. E., Summers, L. H., & Wold, E. G. (2019). Negative nominal interest rates and the bank lending channel, Working Paper No. 25416. National Bureau of Economic Research.

Fehr, E., & Zehnder, C. (2005). *Reputation and credit market formation* (FINRISK Working Paper), University of Zurich.

Farber, V. A., & Reichert, P. (2021). Market infrastructure for social ventures. In O. M. Lehner (Ed.), *A research agenda for social finance*. Edward Elgar Publishing (pp. 79–98).

Fehr, E., & Gächter, S. (2000). Fairness and retaliation: The economics of reciprocity. *Journal of Economic Perspectives, 14*(3): 159–181.

Ferri, G., Kalmi, P., & Kerola, E. (2014). Does bank ownership affect lending behavior? Evidence from the Euro area. *Journal of Banking & Finance, 48*, 194–209.

Ferri, G., & Neuberger, D. (2015). The banking regulatory bubble and how to get out of it. In F. Calciano, F. Fiordelisi, & G. Scarano (Eds.), *The restructuring of banks and financial systems in the Euro Area and the financing of SMEs* (pp. 31–61). Palgrave Macmillan.

GABV. (2020). *Annual report*. Retrieved on January 4, 2022, https://www.gabv.org/wp-content/uploads/2021/12/Annual-Report-2019-EN.pdf

Gafni, H., Hudon, M., & Périlleux, A. (2021). Business or basic needs? The impact of loan purpose on social crowdfunding platforms. *Journal of Business Ethics, 173*(4), 777–793.

Gintis, H., Bowles, S., Boyd, R. T., & Fehr, E. (2004). Moral sentiments and material interests: The foundations of cooperation in economic life. MIT Press.

Godfroid, C., Otiti, N., & Mersland, R. (2022). Employee tenure and staff performance: The case of a social enterprise. *Journal of Business Research, 139*, 457–467.

Gui, B. (1991). The economic rationale for the 'third sector.' *Annals of Public and Cooperative Economics, 62*(4), 551–572.

Guinnane, T. W. (1997). Regional organizations in the German cooperative banking system in the late 19th century. *Research in Economics, 51*(3), 251–274.

Guinnane, T. W. (2001). Cooperatives as information machines: German rural credit cooperatives, 1883–1914. *Journal of Economic History, 61*(2), 366–389.

Guinnane, T. W. (2003). A 'friend and advisor': External auditing and confidence in Germany's credit cooperatives, 1889–1914. *Business History Review, 77*(2), 235–264.

Hesse, H., & Čihák, M. (2007). *Cooperative banks and financial stability* (IMF Working Papers 07/02), Washington: International Monetary Fund.

Karl, M. A. (2015). *Are ethical and social banks less risky? Evidence from a new dataset* (Working Paper 96), DIW Berlin and Institut für Wirtschaftsforschung Halle.

Kalmi, P. (2014). Ethics, banking and ownership. *Bezpieczny Bank, 2*(55), 27–37.

Kitson, A. (1996). Taking the pulse: Ethics and the British cooperative bank. *Journal of Business Ethics, 15*(9), 1021–1031.

Krause, K., & Battenfeld, D. (2019). Coming out of the niche? Social banking in Germany: An empirical analysis of consumer characteristics and market size. *Journal of Business Ethics, 155*(3), 889–911.

Mersland, R., Nyarko, S. A., & Szafarz, A. (2019). Do social enterprises walk the talk? Assessing microfinance performances with mission statements. *Journal of Business Venturing Insights, 11*, e00117.

Meyer, C., & Hudon, M. (2017). Alternative organizations in finance: Commoning in complementary currencies. *Organization, 24*(5), 629–647.

Milano, R. (2011). Social banking: A brief history. In O. Weber & S. Remer (Eds.), *Social banks and the future of sustainable finance* (pp. 15–47). Routledge.

Minetti, R., Murro, P., & Peruzzi, V. (2021). Not all banks are equal. Cooperative banking and income inequality. *Economic Inquiry, 59*(1), 420–440.

Minetti, R., & Zhu, S. C. (2011). Credit constraints and firm export: Microeconomic evidence from Italy. *Journal of International Economics, 83*(2), 109–125.

Morduch, J., & Ogden, T. (2019). The challenges of social investment through the lens of microfinance. In M. Hudon, M. Labie, & A. Szafarz (Eds.), *A research agenda for financial inclusion and microfinance* (pp. 12–26). Edward Elgar Publishing.

Mykhayliv, D., & Zauner, K. G. (2018). The financial and economic performance of social banks. *Applied Economics, 50*(34–35), 3833–3839.

Osterwalder, A., & Pigneur, Y. (2010). *Business model generation: A handbook for visionaries, game changers, and challengers*. Wiley.

Paulet, E., Parnaudeau, M., & Relano, F. (2015). Banking with ethics: Strategic moves and structural changes of the banking industry in the aftermath of the subprime mortgage crisis. *Journal of Business Ethics, 131*(1), 199–207.

Périlleux, A. (2015). When social enterprises engage in finance: Agents of change in lending relationships, a Belgian typology. *Strategic Change, 24*(3), 285–300.

Périlleux, A., Hudon, M., & Bloy, E. (2012). Surplus distribution in microfinance: Differences among cooperative, nonprofit, and shareholder forms of ownership. *Nonprofit and Voluntary Sector Quarterly, 41*(3), 386–404.

Périlleux, A., & Nyssens, M. (2017). Understanding cooperative finance as a new common. *Annals of Public and Cooperative Economics, 88*(2), 155–177.

Postelnicu, L., Hermes, N., & Szafarz, A. (2014). Defining social collateral in microfinance group lending. In R. Mersland & R. Ø. Strøm (Eds.), *Microfinance institutions* (pp. 187–207). Palgrave Macmillan.

Rajan, U., Seru, A., & Vig, V. (2010). Statistical default models and incentives. *American Economic Review, 100*(2), 506–510.

Riedl, A., & Smeets, P. (2017). Why do investors hold socially responsible mutual funds? *Journal of Finance, 72*(6), 2505–2550.

Rizzi, F., Pellegrini, C., & Battaglia, M. (2018). The structuring of social finance: Emerging approaches for supporting environmentally and socially impactful projects. *Journal of Cleaner Production, 170*, 805–817.

San-Jose, L., Retolaza, J. L., & Gutierrez, J. (2011). Are ethical banks different? A comparative analysis using the radical affinity index. *Journal of Business Ethics, 100*(1), 151–173.

Serres, C. (2019). Commons generating companies: What does it take to be one? *Academy of Management Proceedings, 1*, 14352.

Stiglitz, J. E., & Weiss, A. (1981). Credit rationing in markets with imperfect information. *American Economic Review, 71*(3), 393–410.

Tchakoute-Tchuigoua, H. (2010). Is there a difference in performance by the legal status of microfinance institutions? *Quarterly Review of Economics and Finance, 50*(4), 436–442.

Valls Martínez, M. D. C., Cruz Rambaud, S., & Parra Oller, I. M. (2020). Sustainable and conventional banking in Europe. *PloS One, 15*(2), 0229420.

Weber, O. (2013). Social banks and their profitability: Is social banking in line with business success. *Prospective Innovation at Ethical Banking and Finance, 1*(1), 1–19.

Weber, O., & Remer, S. (2011). *Social banks and the future of sustainable finance*. Routledge.
Weber, O., Scholz, R. W., & Michalik, G. (2010). Incorporating sustainability criteria into credit risk management. *Business Strategy and the Environment, 19*(1), 39–50.
Yip, A. W. H., & Bocken, N. M. P. (2018). Sustainable business model archetypes for the banking industry. *Journal of Cleaner Production, 17*, 150–169.

ESG Issues as Strategic Components of Long-term Success of Financial Institutions: Are There Differences in Financial Performance and Firm Value?

Olaf Weber

1 Introduction

This chapter will discuss ESG (Environmental, Social, and Governance) issues as strategic success factors of financial institutions. However, we will mainly focus on banks as major financial institutions. In addition, we will use the terms ESG, corporate sustainability, and corporate social responsibility (CSR) in similar ways, since they all address environmental, social, and governance aspects.

The chapter starts with a historical overview about ESG in the banking industry. Then, we will discuss theoretical explanations for the ESG issues as strategic components of long-term success. This discussion will be followed by an analysis of financial aspects of ESG integration in different banking products and services, such as credit risk assessment,

O. Weber (✉)
University of Waterloo, Waterloo, ON, Canada
e-mail: oweber@uwaterloo.ca

C. Gaganis et al. (eds.), *Sustainable Finance and ESG*, Palgrave Macmillan Studies in Banking and Financial Institutions,
https://doi.org/10.1007/978-3-031-24283-0_2

green lending, and others. Finally, we will focus on some general principles of sustainable banking and how they could be used as a strategic business approach.

2 Historical Overview

The consideration of ESG and sustainability has a relatively long tradition in the banking industry. Modern approaches go back to Italian banks in the sixteenth century. They were founded as an intermediate for those who could save money and those who needed funding for starting or running a business that were needed in the region, such as construction-related trades.

At this time, the lending business often was conducted by loan sharks using usury (Milano, 2011). In contrast, the Italian banks founded at this time were connected to the Catholic Church and therefore judged usury as unethical. Because some of these banks still exist, these were early examples of how ESG criteria positively influenced the business of banks. In addition, these banks also included basic assessment criteria, such as the work ethics of the business owners, their responsibility and efficiency, and their risk-taking (Weber & Feltmate, 2016). Again, we see early approaches of addressing ESG criteria in credit risk assessment that are used by modern banks as well (Weber et al., 2010).

Also, credit unions and cooperatives that were founded in the 1850s in Germany were based on ethical principles. Today, it would be called stakeholder management (Berman et al., 1999). Stakeholder management is often associated with higher firm financial performance (Berman et al., 1999; Freeman, 1984; Scholtens & Zhou, 2008). Their clients were also their members and owners, providing them with democratic participation and financial benefits coming from the revenues of the credit unions. In cities, their members were mainly small and middle-sized company owners while on the country, they mainly served farmers and connected industries. Another difference to other banks with regard to ESG is that credit unions and financial cooperatives have been non-profit organizations. Profits either go to their membership, being lenders, savers, or political units such as municipalities or counties. They addressed the new middle class as well as entrepreneurs and farmers. Credit Unions are still strong in many countries, such as Canada (Desjardins) and Germany. Also, agricultural cooperatives such as Raiffeisen are still very popular in rural regions in Europe. They mainly follow the same ethical principles of their founders. According to the World Council of Credit Unions (https://www.woccu.org/documents/2020_Statistical_Report), they served nearly 400 million members in 2020. Because of their ethical principles, such as focusing on regional activities, and the focus on internalizing the business benefits credit unions suffered much less from the 2008 financial crisis (Li & van Rijn, 2022). Again, the example demonstrates a positive impact of an ESG-oriented business approach, since the cooperative and regional principles avoid credit losses.

Ethical banks founded in the 1960s did not just add ESG indicators to their business but they made them the centre of it. Because of political changes in the 1960s, including social movements and increasing strengths of unions, social criteria, such as financed projects and businesses having a positive societal impact have been considered an essential part of all banking activities of social banks. Furthermore, the E component of ESG, has been emphasized since and the beginning of the discussions around business and the environment caused, among others, by Rachel Carson's book "Silent Spring" (Carson, 2002). Social banks implemented environmental approaches at the core of their business and want to contribute to environmental and societal change. Consequently, they exclusively finance organic farming or the processing of organic products. Hence, the Global Alliance for Banking on Values, an association of 72 ethical banks ion 2022, described this ESF-type of approach as "We put finance at the service of people and the planet" (www.gabv.org). Though social banks are still small and responsible only for a small part of global banking, their number, and assets under management are increasing. Also, their growth is stronger than that of conventional banks and they suffered lees under the 2008 financial crisis (Weber & Feltmate, 2016).

Assessment of ESG Risks in Lending

Lending has been one of the first conventional banking businesses that considered ESG aspects. It has been introduced to assess mainly environmentally induced credit risks and leads to a decrease in credit defaults (Weber et al., 2010, 2015). It has been introduced because of the implementation of environmental regulations in Europe and North America that follow the polluter pays principle. This led to financial risks for lenders of polluters as well (Weber, Fenchel et al., 2008). Later, the approach has been used in other parts of the world as well (Weber et al., 2015). ESG risk assessment is used to manage risks of contaminated sites, used as collateral, costs for borrowers because of environmental regulations, and market changes because of changes in environmental and social attitudes of clients. These risks led to the development of ESG- and sustainability-related credit risk assessment tools that the contribution of environmental, social, and governance factors of the credit risk of commercial loans (Weber, Fenchel et al., 2008; Weber et al., 2010).

ESG-Related Investment

Since the 1990s, ESG criteria also have been used for selecting green, social, and sustainable investments. ESG criteria are used to conduct (socially) responsible investments (RI). Also indexes, such as the Dow Jones Sustainability Indexes, use ESG criteria to select their constituents (Weber & Feltmate, 2016).

The literature on the financial performance of RI compared to conventional investments is vast. Most of the studies found an overperformance or at least the same performance for RI products as the review study by Friede et al. (2015) demonstrates. They also found that the positive ESG impact on the financial performance of companies and consequently for RI products and services is stable. In addition, Weber et al. (2011) found that ESG-based mutual funds outperform their conventional peers in times of turmoil, such as financial crises.

Banks and Climate Risks

At least since the COP21 meeting in Paris in 2015, climate change is on the radar of banks as both, a main financial risk and opportunity. However, historically, climate finance has already been introduced in 1997 with the Kyoto Protocol. The Clean Development Mechanism (CDM) allowed countries to fund projects that reduce GHGs in developing countries abroad to earn carbon credits (Pfaff et al., 2000). Consequently, many banks were lender to the projects. However, in 2012 the CDM mechanism has been out-phased. Since then no UN based official climate-finance mechanisms exist. However, climate finance is still a major part of green finance, including climate bonds.

In 2021, $500 billion have been issued in green bonds. Financially, these bonds are attractive because they usually offer the same financial return as a comparable conventional bond, but in addition they offer a green premium. Many investors ask for the green premium because they want to reduce the climate exposure of their portfolio to reduce climate-related financial risks (Battiston et al., 2021). These risks are increasing for lenders as well (Battiston et al., 2017). Consequently, banks use ESG criteria in their credit assessment processes to reduce these risks. Again, ESG is related to the financial performance of banks and consequently to their firm value.

3 ESG Criteria in Commercial Lending—Financial Aspects

Academic research as well as other reports have demonstrated that ESG performance and financial performance correlate positively (Friede et al., 2015; Klassen & McLaughlin, 1996; Nakao et al., 2007; Weber, 2017). Theoretically, there are a number of explanations for this phenomenon. Some of the prominent theories are institutional theory, slack theory, and good management theory.

Theoretical Background

The relationship between ESG and financial performance is the subject of many studies (Chollet & Sandwidi, 2018; Clark & Viehs, 2014; de Bakker et al., 2005; El Ghoul & Karoui, 2017; Flammer, 2015; Goss & Roberts, 2011; Yannan et al., 2021) and in meta studies (Friede et al., 2015). Though the majority of the studies found a positive correlation between ESG performance and financial performance, it is still open how the connection can be explained theoretically. Often used theories are the slack resources theory, good management theory, stakeholder theory, and institutional theory.

The slack resources theory (Daniel et al., 2004) states that a part of slack resources from financial revenues are used to invest into ESG performance reactively. Consequently, better financial performance leads to better ESG performance. Income is the driver of ESG performance.

In contrast, good management theory (McGuire et al., 1988) claims that ESG management is a part of good management. Therefore, ESG increases financial performance. In this case, ESG is the driver for financial performance. However, often both directions can be found in the literature. Therefore, (Waddock & Graves, 1997) used institutional theory to explain the bi-directional causation.

Also, the resource based view (Wernerfelt, 1984) is closely related to the good management theory. It claims that CSR can have a positive influence on the financial performance because it helps to reduce environmental and social costs, to address stakeholder needs, and to increase a firm's reputation (Deephouse et al., 2016; Lankoski, 2008). Consequently, firms use resources proactively to achieve a competitive advantage through ESG performance (Sharma & Vredenburg, 1998).

Both slack resources and good management theory can also influence each other. Waddock and Graves (1997) call this a bi-directional causality or a virtuous circle. Slack resources in the form of financial assets might be a reason for improved ESG performance. In turn, improved ESG performance might create better financial performance through reputational, costs, and stakeholder effects.

Also, Institutional theory (DiMaggio & Powell, 1983) has been used to explain the bi-directional causality between corporate sustainability and financial performance that is often found in the literature and cannot be explained by both slack resources theory and good management theory (Ameer & Othman, 2012). Institutional theory considers the processes by which structures, including schemas, rules, norms, and routines, become established as authoritative guidelines for social behavior. Furthermore, it explains how these elements are created, diffused, adopted, and adapted (Scott, 1987). Consequently, also firms are influenced by coercive (regulative), normative (social norms), or mimetic (mimicking competitors) pressure. This pressure might also cause an increase in ESG performance. An example for coercive pressure is the introduction of environmental regulations, such as the Chinese Green Credit Guidelines. These regulations increased the ESG and the financial performance of Chinese banks (Cui et al., 2018). Voluntary codes of conduct are an example for normative pressure. If a firm becomes a member of such a voluntary code, for instance, the Equator Principles for Project Finance. A study by Weber (2016) showed that the ESG reporting quality increases if financial institutions are members of the Equator Principles. Mimetic pressure appears if competitors increase their ESG activities successfully and others will imitate this successful behavior.

To summarize, there is a number of theories that explain the connection between ESG performance and financial performance. The explanations, however, are manifold and it needs detailed research to explain the connection for a specific firm or industry. Also, the quality of the ESG performance data is rather low compared to financial data because it is not mandatory and is often incomplete. This makes it harder to analyze and to explain the connection between ESG and financial performance.

4 ESG Integration in Financial Products and Services

ESG Investing

ESG Investing also called Responsible Investing (RI) and Socially Responsible Investing (SRI), has left its niche and became mainstream. The Global Sustainable Investment Alliance estimates that ESG-based assets under management reached USD35.3 trillion in 2020, with a growth of 15% in two years (Global Sustainable Investment Alliance, 2021). Overall, they state that ESG Investing counts for 35% of all investments.

To stay consistent, we use the term ESG Investing to describe the following type of investing "...the integration of environmental, social and governance (ESG) factors in the selection and management of investments" (Bragg & Smeh, 2013). Major socially responsible investing (SRI) strategies that can be identified are positive and negative screening, the integration of ESG factors in investment decisions, sustainability themed investing, and corporate engagement and shareholder action. We address Impact Investing in a separate section outside of ESG screening since it follows a different goal than ESG investing.

Generally, ESG Investing has two main goals. First, ESG wants to guarantee attractive financial returns by investing in securities that take long-term sustainability concerns into account. Second, ESG Investing wants to shift capital toward activities that have a positive social, environmental, or sustainability benefit (or a less negative sustainability impact), and therefore support a sustainable development (Weber & Feltmate, 2016). The rationale behind the first goal is that firms and other investments that address ESG criteria perform better financially. This logic and the theories that explain it are described above in this chapter. The rationale behind the second goal is that sometimes investors do not only strive for profit maximization but for positive environmental and societal impacts as well. However, ESG Investing does not guarantee that the investments and consequently the investment portfolio becomes greener or more social.

ESG Investment basically means that ESG criteria are considered in investment decisions. Nevertheless, considering does not mean that investment will be made or not made based on ESG criteria. Investors might acknowledge the good or bad ESG performance of an investment but make or do not make the investment anyway. This approach of "integrating" ESG criteria is also one of the reasons that ESG Investments are estimated that high (see above). Many portfolios that are based on an ESG approach are not really different from conventional portfolios. This issue has even led to regulatory action against investors claiming that their funds follow an ESG approach. Some investors are accused for greenwashing with regard to their sustainable investment products (Heitzner, 2022).

With regard to the financial success of ESG Investments, studies and meta-studies found that ESG Investments perform similar as conventional investments, and that many ESG Investments even outperform their conventional counterparts (Busch & Friede, 2018; Friede et al., 2015; Weber & Ang, 2016; Weber, Koellner, et al., 2008). Often, it is found that these investments outperform their conventional counterparts in times of crises (Weber et al., 2011). Some studies, however, could not find this characteristic during the COVID pandemic (Folger-Laronde et al., 2020). Additional factors, such as the general portfolio management and the investment decision-making, regional and sectoral factors, and the type of ESG indicators used, might be variables that explain the variance in the performance of ESG Investment.

Impact Investing

Impact investment is a form of investment that addresses social or environmental challenges and generates financial returns. In contrast to ESG Investing, the creation of positive impacts is a necessity in Impact Investing. Usually, societal impacts have a higher priority than financial returns. However, the spectrum of financial returns as well as impact vary. Therefore, some impact investments may create financial returns that are comparable to conventional investments. For instance, Porter and Kramer (2011) describe such an approach in his article about the shared value approach. They claim that investments addressing societal needs are also more successful from a financial point of view. However, Impact Investing is similar to ESG Investing as its principles are diluted.

Busch et al. (2021) claim that impact investing originally focused on achieving transformational changes. Nowadays, however, the term is used interchangeably for any investments that consider ESG criteria. Consequently, they ask for metrics that can be used to measure the impact of the investment. Such metrics could be, GHG emissions reduced by $X of investment, or number of schools built by $X of investment. Often, however, these metrics are not measured though systems, such as IRIS by the Global Impact Investing Network provides a system of indicators to track the social and environmental impact of Impact Investments.

Financially, Impact Investments might outperform other types of investments or not. Often, this depends on the choice of the investors and on the type of investment. Because, the creation of a societal transformation is the first goal of an Impact Investor, the financial return might be less important. There might be cases, such as investments in social enterprises that create high returns, but other cases might not create such a win-win situation. In some cases, Impact Investors even abstain from a financial return at all or even provide their investment as a grant.

The attitude toward the ratio between impact and financial return might also depend on the type of impact investors. Specialized impact investors might strive for more societal impact. The same might be true for foundations that focus on specific impact investment topics. The ratio might be different for investors that conduct impact investment as a small part of their financial activities. Some banks, for instance, conduct impact investment with a small part of their assets and consequently might strive to returns comparable to conventional investments. However, even specialized impact investors often achieve financial returns comparable to conventional investments.

Fossil Fuel Divestment and Engagement

Analyses of the Intergovernmental Panel on Climate Change (IPCC) state that only a small part of the fossil fuel reserves that are still in the ground can be burned if the world wants to stay below a 2 °C warming (IPCC, 2021). Among others, this has serious consequences for investments in the fossil fuel industry. Hence, there are both moral and financial reasons to rethink investments in the fossil fuel industry.

Firstly, the fossil fuel industry may be tagged as "immoral" because fossil fuel production and consumption contributes to two-thirds of the CO_2 emissions that cause climate change (Ekwurzel et al., 2017; Heede & Oreskes, 2016). Consequently, some investors might divest from the industry because they do not want to support firms that cause climate change. Financially, even announcement for such decisions have a negative effect on the share price of firms in the fossil fuel industry (Dordi & Weber, 2019) as well as a positive effect on the share price of the divesting investor (Bassen et al., 2020). Furthermore, studies demonstrated that portfolios that are less exposed to climate change perform better financially (Henriques & Sadorsky, 2018; Hunt & Weber, 2019; Trinks et al., 2018). This leads to the second reasons to divest from or to engage with the fossil fuel industry, the financial motivation.

Secondly, a number of studies even found that investments in the fossil-fuel sector create financial risks for investors (Battiston et al., 2017; Monasterolo & De Angelis, 2019; Monasterolo et al., 2017) because of stranded assets (Ansar et al., 2014; Green & Newman, 2017) caused by the limited opportunities to burn fossil fuel reserves (Campiglio et al., 2018, 2019), and because of political decisions to transition to a low-carbon economy might have negative financial impacts on the fossil fuel industry (Linnenluecke et al., 2015; Strauch et al., 2020).

Consequently, investors might reduce their investments in the industry or engage with their investees to make their business in-line with climate needs (O'Rourke, 2003; Othman & Ameer, 2010; Schaltegger & Burritt, 2015). Such investment strategies might have a strong impact on the fossil fuel industry and climate change because only a small number of investors own a significant portion of the fossil fuel industry, and consequently, are able to apply pressure on their investees (Dordi et al., 2022).

Green Lending

Green lending exists until the 1990s (Weber & Feltmate, 2016). The rationale is that lending to green borrowers including Greentech is a good business opportunity because these borrowers address a market demand. Consequently, green loans have a lower default probability than non-green borrowers. The same approach has been used for mortgages Green housed might have lower energy costs, have a higher value and—according to the good management theory (Waddock & Graves, 1997)—their owner conduct a better financial management.

Recently, green lending increased again because of government policies and incentives. For instance, the European Union and the European Banking Federation have issued green and sustainable finance guidelines (Cui et al., 2018). Furthermore, members of the Sustainable Banking Network hosted by the International Finance Corporation (IFC), have introduced green and sustainable financial regulations. Out of these, the Chinese Green Credit Guidelines are probably the most prominent green credit guidelines (Aizawa & Chaofei, 2010; Cheng et al., 2021; China Banking Regulatory Commission, 2012, 2014; Zhang et al., 2011; C. Zhao, 2015; N. Zhao & Xu, 2012). They ask for a shift of the lending portfolios away from polluting industries to green industries. Consequently, the Chinese financial regulators have developed indicators that measure the progress of lenders with regard to green lending. Banks have to demonstrate that they increase the ratio of green loans in the credit portfolio. Overall, it seems that green lending in China is also successful from a financial point of view (Cui et al., 2018).

Bangladesh Bank uses an incentive-based approach to channel loans to green or less polluting industries. They provide lenders with lower interest loans if they apply an environmental credit risk assessment scheme and consequently prefer greener borrower (Bangladesh Bank, 2011). Again, it looks like using an environmental risk management tool in commercial lending decreases the probability of default of loans (Weber et al., 2015). However, banks need to follow a more proactive approach with regard to green lending to increase the benefit of green lending (Weber & Chowdury, 2020).

ESG-Related Credit Risk Assessment

Similar to ESG Investing, ESG-related credit risk management uses ESG criteria in addition to financial criteria to manage credit risks. The risks might be environmental, societal, and climate-related. With an increasing climate emergency and the introduction of GHG emission pricing, the need to address climate-related credit risks increases.

Battiston et al. (2017), for instance, analyzed the influence of climate exposure on portfolio risks. They found higher risks for lending portfolios that are more exposed to climate risk. Other studies analyzed the connection between GHG emissions and the credit default probability of commercial borrowers. Monnin (2018) showed increased probability of default for borrowers from the Utilities and the Material sectors. Bouchet and Le Guenedal (2020) also found significant EBITDA losses and consequently and increase in the probability of default for firms in the Energy, Materials, and Utilities sectors. Also, Capasso et al. (2020) found a negative effect of carbon emissions of the creditworthiness that is amplified by climate-related events, such as the Paris agreement. Finally, Oyegunle et al. (2022) found a negative impact of a price on carbon emissions on the credit default probability.

Also, studies that used a broader sustainability approach found a connection between ESG performance and creditworthiness (Bauer & Hann, 2010; Höck et al., 2020; Weber, 2012; Weber et al., 2010). They all state that it makes financial sense to consider ESG criteria in lending decisions to avoid ESG-related default risks.

Green Bonds

To date, green bonds are single biggest source of capital with a higher amount than equity (Weber & Saravade, 2019). Green bonds are issued to raise long-term debt capital from various domestic and international investors to either finance or refinance green assets and projects (Saravade & Weber, 2020). Hence, their use of proceeds goes toward green projects and assets. They are issued by national and regional governments, financial institutions, such as banks and multilateral development banks (MDB), and by firms that want to finance ESG activities. Since its inception in 2007, the frequency of issuances and the financial value of green and climate bond issuances have increased significantly. In 2021, the global market reached $1.2 trillion of cumulative issuances (Climate Bonds Initiative, 2022). Consequently, green bonds are an important financial product that helps to meet the 1.5 °C target.

Similar to conventional bonds, eight types of green bonds exist (Weber et al., 2018). Corporate bonds, usually use-of-proceeds bond, are backed by the issuing corporation's balance sheet. Secondly, project bonds are backed by earnings of a single project or multiple projects. To disburse the proceeds of these projects special purpose vehicles (SPV) that are independent subsidiaries of the issuer, are established. They disconnect the risk of the project bond from the issuing corporation. Thirdly, multiple projects, such as windfarms or photovoltaic projects, might be grouped and collateralized to create an asset-backed security (ABS). Fourthly, covered bonds are secured with underlying assets to cover the bond if the issuer defaults.

Fifthly, financial sector bonds are used by the financial industry for "on-balance sheet" lending. Often, the use of proceed is defined broadly to provide the financial institution with the opportunity to use them for a variety of projects and corporations. Sixthly, multilateral institutions, such as the World Bank and other MDBs issue supranational, sub-sovereign and agency bonds. Seventhly, municipal bonds are issued by regional governments, municipalities, and cities. Finally, sovereign green bonds are issued by national governments. Their proceeds go toward green public sector projects, for instance, green infrastructure projects.

Because the interest of green bonds is based on the risk of the issuer, they offer the same financial returns as conventional bonds, but offer an additional green premium (Saravade & Weber, 2020). This makes them attractive for institutional investors, such as pension funds, insurance companies, hedge funds, mutual funds, sovereign wealth funds, and endowments that often need to achieve certain financial returns because of their fiduciary duty but want to have more green investments that decrease the climate risk exposure and increase the environmental performance of their portfolios.

To summarize, the proceeds of green bonds go toward green projects and assets. Financially, green bonds are attractive because they offer fixed returns that depend on the risk of the issuer. In addition to the financial attractiveness, green bonds offer a green premium, also called greenium.

5 Conclusion

This section presented the effect of ESG criteria in financial decision-making. We could show that considering ESG criteria has not only had an environmental and societal effect, but a financial effect as well. Nearly all academic studies agree that considering ESG criteria has a positive effect on financial risk. Furthermore, introducing ESG-related products and services might create additional income as it meets market demands.

Consequently, ESG issues should become strategic components of firms inside and outside the financial sector to guarantee long-term success. Both, investors and lenders use ESG criteria to price their investments and loans. Hence, a borrower with a good ESG performance is more likely to achieve credit and then interest rate of the loan might be lower than that for a low ESG performer. Hence, the investment in ESG wit pay off. It is a so-called sustainability case—better sustainability increases the financial performance (Weber & Feltmate, 2016).

Furthermore, there is a stronger demand that firms prove their positive contribution to society. They need to show that they do not contribute to climate change and other societal challenges, such as discrimination of minorities, including indigenous people. Banks and other investors, for instance, are criticized because they still finance fossil fuel companies and consequently contribute to climate change (Dordi et al., 2022). This has a negative effect on their reputation and might lead to long-term financial underperformance. Integrating ESG into the long-term business strategy might help to avoid this problem.

Integrating ESG into the corporate business strategy, however, is not without difficulties. Currently, ESG criteria are still not standardized, and the quality of the indicators is often hard to evaluate. This might also be one of the reasons that studies about the correlation between ESG and financial performance deliver mixed results. Furthermore, ESG expertise is needed that is often not available in corporations. Here, the need to integrate ESG knowledge into business school education is obvious. However, with an increase of environmental and societal challenges, the benefits of addressing ESG issues strategically are obvious.

References

Aizawa, M., & Chaofei, Y. (2010). Green credit, green stimulus, green revolution? China's mobilization of banks for environmental cleanup. *The Journal of Environment & Development, 19*(2), 119–144.

Ameer, R., & Othman, R. (2012). Sustainability practices and corporate financial performance: A study based on the top global corporations. *Journal of Business Ethics, 108*(1), 61–79. https://doi.org/10.1007/s10551-011-1063-y

Ansar, A., Caldecott, B., & Tilbury, J. (2014). *Stranded assets and the fossil fuel divestment campaign: What does divestment mean for the valuation of fossil fuel assets?* Smith School of Enterprise and the Environment, University of Oxford.

Bangladesh Bank. (2011). *Environmental Risk Management (ERM) guidelines for banks and financial institutions in Bangladesh.* Bangladesh Bank.

Bassen, A., Kaspereit, T., & Buchholz, D. (2020). The capital market impact of Blackrock's thermal coal divestment announcement. *Finance Research Letters*, 101874. https://doi.org/10.1016/j.frl.2020.101874

Battiston, S., Mandel, A., Monasterolo, I., Schuetze, F., & Visentin, G. (2017). A climate stress-test of the financial system. *Nature Climate Change, 7*(4), 283–288. https://doi.org/10.1038/nclimate3255

Battiston, S., Monasterolo, I., Riahi, K., & van Ruijven, B. J. (2021). Accounting for finance is key for climate mitigation pathways. *Science*, eabf3877. https://doi.org/10.1126/science.abf3877

Bauer, R., & Hann, D. (2010). Corporate environmental management and credit risk. *SSRN eLibrary*. http://ssrn.com/paper=1660470

Berman, S. L., Wicks, A. C., Kotha, S., & Jones, T. M. (1999). Does stakeholder orientation matter? The relationship between stakeholder management models and firm financial performance. *The Academy of Management Journal, 42*(5), 488–506. https://doi.org/10.2307/256972

Bouchet, V., & Le Guenedal, T. (2020). *Credit risk sensitivity to carbon price*. https://ssrn.com/abstract=3574486

Bragg, I., & Smeh, D. T. (2013). *Canadian socially responsible investment review 2012*. Retrieved from Toronto.

Busch, T., Bruce-Clark, P., Derwall, J., Eccles, R., Hebb, T., Hoepner, A., Klein, C., Krueger, P., Paetzold, F., Scholtens, B., & Weber, O. (2021). Impact investments: A call for (re)orientation. *SN Business & Economics, 1*(2), 33. https://doi.org/10.1007/s43546-020-00033-6

Busch, T., & Friede, G. (2018). The robustness of the corporate social and financial performance relation: A second-order meta-analysis. *Corporate Social Responsibility and Environmental Management, 25*(4), 583–608. https://doi.org/10.1002/csr.1480

Campiglio, E., Dafermos, Y., Monnin, P., Ryan-Collins, J., Schotten, G., & Tanaka, M. (2018). Climate change challenges for central banks and financial regulators. *Nature Climate Change, 8*(6), 462–468. https://doi.org/10.1038/s41558-018-0175-0

Campiglio, E., Monnin, P., & von Jagow, A. (2019). *Climate risks in financial assets*. CEP Council on Economic Policies.

Capasso, G., Gianfrate, G., & Spinelli, M. (2020). Climate change and credit risk. *Journal of Cleaner Production, 266*, 121634.

Carson, R. (2002). *Silent spring*. Houghton Mifflin Harcourt.

Cheng, Q., Lai, X., Liu, Y., Yang, Z., & Liu, J. (2021). The influence of green credit on China's industrial structure upgrade: Evidence from industrial sector panel data exploration. *Environmental Science and Pollution Research*, 1–15.

China Banking Regulatory Commission. (2012). *Notice of the China banking regulatory commission CBRC on Issuing the green credit guidelines*. China Banking Regulatory Commission.

Chollet, P., & Sandwidi, B. W. (2018). CSR engagement and financial risk: A virtuous circle? International evidence. *Global Finance Journal*. https://doi.org/10.1016/j.gfj.2018.03.004

Clark, G. L., & Viehs, M. (2014). *The implications of corporate social responsibility for investors: An overview and evaluation of the existing CSR literature*. Retrieved from SSRN: http://ssrn.com/abstract=2481877

Climate Bonds Initiative. (2022). *Bonds and climate change: The state of the market 2021.* Climate Bonds Initiative.

Cui, Y., Geobey, S., Weber, O., & Lin, H. (2018). The impact of green lending on credit risk in China. *Sustainability, 10*(6), 2008. https://doi.org/10.3390/su10062008

Daniel, F., Lohrke, F. T., Fornaciari, C. J., & Turner, R. A., Jr. (2004). Slack resources and firm performance: A meta-analysis. *Journal of Business Research, 57*(6), 565–574.

de Bakker, F. G. A., Peter Groenewegen, P., & den Hond, F. (2005). A bibliometric analysis of 30 years of research and theory on corporate social responsibility and corporate social performance. *Business Society, 44*, 283–317.

Deephouse, D. L., Newburry, W., & Soleimani, A. (2016). The effects of institutional development and national culture on cross-national differences in corporate reputation. *Journal of World Business, 51*(3), 463–473. https://doi.org/10.1016/j.jwb.2015.12.005

DiMaggio, P. J., & Powell, W. W. (1983). The Iron Cage revisited: Institutional isomorphism and collective rationality in organizational fields. *American Sociological Review, 48*(2), 147–160.

Dordi, T., Gehricke, S. A., Naef, A., & Weber, O. (2022). Ten financial actors can accelerate a transition away from fossil fuels. *Environmental Innovation and Societal Transitions, 44*, 60–78. https://doi.org/10.1016/j.eist.2022.05.006

Dordi, T., & Weber, O. (2019). The Impact of Divestment announcements on the share price of fossil fuel stocks. *Sustainability, 11*(11), 3122.

Ekwurzel, B., Boneham, J., Dalton, M. W., Heede, R., Mera, R. J., Allen, M. R., & Frumhoff, P. C. (2017). The rise in global atmospheric CO_2, surface temperature, and sea level from emissions traced to major carbon producers. *Climatic Change, 144*(4), 579–590. https://doi.org/10.1007/s10584-017-1978-0

El Ghoul, S., & Karoui, A. (2017). Does corporate social responsibility affect mutual fund performance and flows? *Journal of Banking Finance, 77*, 53–63. https://doi.org/10.1016/j.jbankfin.2016.10.009

Flammer, C. (2015). Does corporate social responsibility lead to superior financial performance? A regression discontinuity approach. *Management Science, 61*(11), 2549–2568. https://doi.org/10.1287/mnsc.2014.2038

Folger-Laronde, Z., Pashang, S., Feor, L., & ElAlfy, A. (2020). ESG ratings and financial performance of exchange-traded funds during the COVID-19 pandemic. *Journal of Sustainable Finance & Investment*, 1–7. https://doi.org/10.1080/20430795.2020.1782814

Freeman, R. E. (1984). *Strategic management: A stakeholder approach.* Prentice-Hall.

Friede, G., Busch, T., & Bassen, A. (2015). ESG and financial performance: Aggregated evidence from more than 2000 empirical studies. *Journal of Sustainable Finance & Investment, 5*(4), 210–233. https://doi.org/10.1080/20430795.2015.1118917

Global Sustainable Investment Alliance. (2021). *2020 global sustainable investment review*. Global Sustainable Investment Alliance.

Goss, A., & Roberts, G. S. (2011). The impact of corporate social responsibility on the cost of bank loans. *Journal of Banking and Finance, 35*(7), 1794–1810.

Green, J., & Newman, P. (2017). Disruptive innovation, stranded assets and forecasting: The rise and rise of renewable energy. *Journal of Sustainable Finance & Investment, 7*(2), 169–187. https://doi.org/10.1080/20430795.2016.1265410

Heede, R., & Oreskes, N. (2016). Potential emissions of CO_2 and methane from proved reserves of fossil fuels: An alternative analysis. *Global Environmental Change, 36*, 12–20. https://doi.org/10.1016/j.gloenvcha.2015.10.005

Heitzner, C. (2022, May 31). Deutsche Bank raided by authorities over ESG 'greenwashing' claims: 'We've found evidence that that could support allegations of prospectus fraud'. *Fortune*. https://fortune.com/2022/05/31/deutsche-bank-dws-esg-greenwashing-raid-evidence-seized-whistleblower-fixler/#:~:text=%E2%80%9CThe%20allegations%20are%20that%20DWS,into%20the%20claims%20since%20January.

Henriques, I., & Sadorsky, P. (2018). Investor implications of divesting from fossil fuels. *Global Finance Journal, 38*, 30–44. https://doi.org/10.1016/j.gfj.2017.10.004

Höck, A., Klein, C., Landau, A., & Zwergel, B. (2020). The effect of environmental sustainability on credit risk. *Journal of Asset Management, 21*(2), 85–93. https://doi.org/10.1057/s41260-020-00155-4

Hunt, C., & Weber, O. (2019). Fossil fuel divestment strategies: Financial and carbon related consequences. *Organization & Environment, 32*(1), 41–61. https://doi.org/10.1177/1086026618773985

IPCC. (2021). *Climate change 2021: The physical science basis. Contribution of working group I to the sixth assessment report of the intergovernmental panel on climate change*. Cambridge University Press.

Klassen, R. D., & McLaughlin, C. P. (1996). The impact of environmental management on firm performance. *Management Science, 42*(8), 1199–1214.

Lankoski, L. (2008). Corporate responsibility activities and economic performance: A theory of why and how they are connected. *Business Strategy and the Environment, 17*(8), 536–547. https://doi.org/10.1002/bse.582

Li, K., & van Rijn, J. (2022). Credit union and bank subprime lending in the great recession. *The Review of Corporate Finance Studies*.

Linnenluecke, M. K., Meath, C., Rekker, S., Sidhu, B. K., & Smith, T. (2015). Divestment from fossil fuel companies: Confluence between policy and strategic viewpoints. *Australian Journal of Management, 40*(3), 478–487. https://doi.org/10.1177/0312896215569794

McGuire, J. B., Sundgren, A., & Schneeweis, T. (1988). Corporate social responsibility and firm financial performance. *Academy of Management Journal, 31*(4), 854–872.

Milano, R. (2011). Social banking: A brief history. In O. Weber & S. Remer (Eds.), *Social banks and the future of sustainable finance* (Vol. 64, pp. 15–47). Routledge.

Monasterolo, I., Battiston, S., Janetos, A. C., & Zheng, Z. (2017). Vulnerable yet relevant: The two dimensions of climate-related financial disclosure. *Climatic Change, 145*(3), 495–507. https://doi.org/10.1007/s10584-017-2095-9

Monasterolo, I., & De Angelis, L. (2019). Blind to carbon risk? An analysis of stock market's reaction to the Paris Agreement. *Ecological Economics, 170*, 106571.

Monnin, P. (2018). *Integrating climate risks into credit risk assessment-current methodologies and the case of central banks corporate bond purchases* (Council on Economic Policies, Discussion Note, 4).

Nakao, Y., Amano, A., Matsumura, K., Genba, K., & Nakano, M. (2007). Relationship between environmental performance and financial performance: An empirical analysis of Japanese corporations. *Business Strategy and the Environment, 16*, 106–118.

Notice of the China Banking Regulatory Commission on Key Performance Indicators of Green Credit Implementation, 186 C.F.R. (2014).

O'Rourke, A. (2003). A new politics of engagement: Shareholder activism for corporate social responsibility. *Business Strategy and the Environment, 12*(4), 227–239. https://doi.org/10.1002/bse.364

Othman, R., & Ameer, R. (2010). Environmental disclosures of palm oil plantation companies in Malaysia: A tool for stakeholder engagement. *Corporate Social Responsibility and Environmental Management, 17*(1), 52–62. https://doi.org/10.1002/csr.218

Oyegunle, A., Weber, O., & ElAlfy, A. (2022). Carbon costs and credit risk in a resource-based economy—A scenario analysis. *Journal of Sustanable Finance & Investment*.

Pfaff, A.S., Kerr, S., Hughes, R. F., Liu, S., Sanchez–Azofeifa, G. A., Schimel, D., Tosi, J., & Watson, V. (2000). The Kyoto protocol and payments for tropical forest: An interdisciplinary method for estimating carbon-offset supply and increasing the feasibility of a carbon market under the CDM. *Ecological Economics, 35*(2), 203–221.

Porter, M. E., & Kramer, M. R. (2011). Creating shared value. *Harvard Business Review, 89*(1–2), 62–77. http://search.ebscohost.com/login.aspx?direct=true&db=bth&AN=56698455&site=ehost-live&scope=site

Saravade, V., & Weber, O. (2020). An institutional pressure and adaptive capacity framework for green bonds: Insights from India's emerging green bond market. *World, 1*(3), 239–263. https://www.mdpi.com/2673-4060/1/3/18

Schaltegger, S., & Burritt, R. (2015). Business cases and corporate engagement with sustainability: Differentiating ethical motivations. *Journal of Business Ethics*, 1–19. https://doi.org/10.1007/s10551-015-2938-0

Scholtens, B., & Zhou, Y. (2008). Stakeholder relations and financial performance. *Sustainable Development, 16*(3), 213–232. https://doi.org/10.1002/sd.364

Scott, W. R. (1987). The adolescence of institutional theory. *Administrative Science Quarterly, 32*(4), 493–511. http://www.jstor.org/stable/2392880

Sharma, S., & Vredenburg, H. (1998). Proactive corporate environmental strategy and the development of competitively valuable organizational capabilities. *Strategic Management Journal, 19*, 729–753.

Strauch, Y., Dordi, T., & Carter, A. (2020). Constraining fossil fuels based on 2 °C carbon budgets: The rapid adoption of a transformative concept in politics and finance. *Climatic Change, 160*(2), 181–201. https://doi.org/10.1007/s10584-020-02695-5

Trinks, A., Scholtens, B., Mulder, M., & Dam, L. (2018). Fossil fuel divestment and portfolio performance. *Ecological Economics, 146*, 740–748. https://doi.org/10.1016/j.ecolecon.2017.11.036

Waddock, S., & Graves, S. B. (1997). The corporate social performance—Financial performance link. *Strategic Management Journal, 18*(4), 303–319. https://doi.org/10.1002/(sici)1097-0266(199704)18:4%3c303::aid-smj869%3e3.0.co;2-g

Weber, O. (2012). Environmental credit risk management in banks and financial service institutions. *Business Strategy and the Environment, 21*(4), 248–263. https://doi.org/10.1002/bse.737

Weber, O. (2016). Equator principles reporting: Factors influencing the quality of reports. *International Journal of Corporate Strategy and Social Responsibility, 1*(2), 141–160.

Weber, O. (2017). Corporate sustainability and financial performance of Chinese banks. *Sustainability Accounting, Management and Policy Journal, 8*(3), 358–385. https://doi.org/10.1108/SAMPJ-09-2016-0066

Weber, O., & Ang, W. R. (2016). The performance, volatility, persistence and downside risk characteristics of sustainable investments in emerging market. *ACRN Oxford Journal of Finance and Risk Perspectives, 5*(2), 1–13.

Weber, O., & Chowdury, R. K. (2020). Corporate sustainability in Bangladeshi banks: Proactive or reactive ethical behavior? *Sustainability, 12*(19), 7999. https://www.mdpi.com/2071-1050/12/19/7999

Weber, O., & Feltmate, B. (2016). *Sustainable banking: Managing the social and environmental impact of financial institutions*. University of Toronto Press.
Weber, O., Fenchel, M., & Scholz, R. W. (2008). Empirical analysis of the integration of environmental risks into the credit risk management process of European banks. *Business Strategy and the Environment, 17*, 149–159. https://doi.org/10.1002/bse.507
Weber, O., Hoque, A., & Islam, A. M. (2015). Incorporating environmental criteria into credit risk management in Bangladeshi banks. *Journal of Sustainable Finance & Investment, 5*(1–2), 1–15. https://doi.org/10.1080/20430795.2015.1008736
Weber, O., Koellner, T., Habegger, D., Steffensen, H., & Ohnemus, P. (2008). The relation between sustainability performance and financial performance of firms. *Progress in Industrial Ecology, 5*(3), 236–254.
Weber, O., Mansfeld, M., & Schirrmann, E. (2011). The financial performance of RI funds after 2000. In W. Vandekerckhove, J. Leys, K. Alm, B. Scholtens, S. Signori, & H. Schaefer (Eds.), *Responsible investment in times of turmoil* (pp. 75–91). Springer.
Weber, O., & Saravade, V. (2019). Green bonds—Current development and their future. In (p. 32). Centre for International Governance Innovation.
Weber, O., Saravade, V., & Dordi, T. (2018). *Strategies and policies for integrating the Canadian financial sector into financing the transition to a low-carbon economy*. School of Environment, Enterprise and Development.
Weber, O., Scholz, R. W., & Michalik, G. (2010). Incorporating sustainability criteria into credit risk management. *Business Strategy and the Environment, 19*(1), 39–50. https://doi.org/10.1002/bse.636
Wernerfelt, B. (1984). A resource-based view of the firm. *Strategic Management Journal, 5*(2), 171–180. https://doi.org/10.2307/2486175
Yannan, D., Ahmed, A. A. A., Kuo, T. H., Malik, H. A., Nassani, A. A., Haffar, M., Suksatan, W., & Iramofu, D. P. F. (2021). Impact of CSR, innovation, and green investment on sales growth: new evidence from manufacturing industries of China and Saudi Arabia. *Economic Research-Ekonomska Istraživanja*, 1–20. https://doi.org/10.1080/1331677X.2021.2015610
Zhang, B., Yang, Y., & Bi, J. (2011). Tracking the implementation of green credit policy in China: Top-down perspective and bottom-up reform. *Journal of Environmental Management, 92*(4), 1321–1327. https://doi.org/10.1016/j.jenvman.2010.12.019
Zhao, C. (2015). Commercial banks' green credit practice as a support for the green economic transformation (transl.). *Finance and Accounting Monthly, 32*, 23.
Zhao, N., & Xu, X.-J. (2012). Analysis on green credit in China. *Advances in Applied Economics and Finance (AAEF), 3*(21), 501–506.

Corporate Social Responsibility, Customer Satisfaction, and Customer Loyalty in Banking Institutions: A Literature Review

Stratos Kartsonakis and Evangelos Grigoroudis

1 Introduction

Corporate social responsibility (CSR) is a concept that is constantly being developed and its aim is to affect positively both customers and stakeholders (Zhang, 2020). CSR has been established as a major research field in academia, but it has also gained significant role in industries and societies, due to the fact that it includes various dimensions, such as economic, philanthropic, ethical, and legal factors (Park & Kim, 2019). Nowadays firms are expected to produce profit for the stakeholders without disregarding a socially responsible behavior (Aramburu & Pescador, 2019). The recent economic recession indicated that overfocusing on generating greater financial results, without considering other

S. Kartsonakis · E. Grigoroudis (✉)
School of Production Engineering and Management, Technical University of Crete, Chania, Greece
e-mail: egrigoroudis@tuc.gr

S. Kartsonakis
e-mail: ekartsonakis@tuc.gr

C. Gaganis et al. (eds.), *Sustainable Finance and ESG*,
Palgrave Macmillan Studies in Banking and Financial Institutions,
https://doi.org/10.1007/978-3-031-24283-0_3

business-related factors may lead to failure. Such failures can negatively affect the stability of financial systems and have severe social and environmental consequences (Platonova et al., 2018). To avoid such impacts, it is of great importance for firms and institutions to alter their corporate governance which will be more considerate to the environment and the society they operate by prioritizing CSR activities.

The concept of CSR was introduced in the literature in the early 1950s, but it did not start being adopted by practitioners until the late 1990s as an important part of their business practices (Carroll, 1999). The development of CSR activities by industries demands, however, large amounts of effort and money which will improve both tangible and intangible attributes, such as customer loyalty, better brand image, and increased reputation, as well as better financial performance (Mcdonald, 2007). More specifically, Weber (2008) describes the benefits of CSR as follows:

- Increasing employee motivation
- Retention and recruitment
- Enhanced revenues
- Improving brand image and reputation.

Moreover, Carroll (2016) suggested that CSR may offer extra value to industries by meeting expectations for good governance policies, new opportunities for growth, and improving access to innovation and customer/employee engagement. It should be noticed though, that empirical results are not always leading to these outcomes. Nonetheless, in order for firms to be able to take advantage of the CSR benefits, there should be a clear understanding of the CSR practices that are being implemented (i.e., how each firm defines and applies CSR) and the expectations that the shareholders have (Pérez & Rodríguez, 2015a, 2015b). Under this context, the allocation of their resources should be made in ways that bring the optimum benefits for both societies and stakeholders, such as customers, employees, supply chain partners, and governments (Poolthong, 2009).

A sector that over the last years is incorporating CSR practices in its operational and business practices is the banking industry (Platonova et al., 2018). The banking sector nowadays has an important role in modern societies as it not only functions as a financial stability factor for economies, but it paves the way for new trends and strategies and provides multiple services to customers, and thus it is expected to be more socially responsible (Chambers & Day, 2009). Moreover, the banking industry, especially in the recent years, has experienced various transformations, such as the latest economic recession, globalization, financial innovation, and the emergence of new technologies which are affecting the distribution of bank services (Flavia & Torres, 2005). As a result, society has lost its confidence to the financial institutions and the various shareholders (customers, employees, etc.) are demanding better tools for the evaluation of the banking sector (Pérez & Rodríguez, 2015a, 2015b). Under this context, banks adopt CSR practices in order to improve their corporate image and their customers' confidence (Flavia & Torres, 2005). In this scope, practitioners and academics have researched the importance of CSR practices in the banking sector.

The aim of this chapter is to study the existing literature and examine whether social and environmental responsibility in the banking and mutual funds institutions can enhance customer satisfaction and customer loyalty. The rest of this chapter is organized as follows: it begins with an analysis of the various definitions and theories around the CSR, as well as the definitions of customer satisfaction and customer loyalty. The next section presents the research that have been made regarding CSR in the banking sector, while fourth section is dedicated to the literature related to mutual funds. Finally, a discussion of the literature findings is presented, and the chapter concludes by discussing limitations and suggestions for future research.

2 Corporate Social Responsibility (CSR)

As already mentioned, CSR is a concept that has received a lot of attention in the literature, but it still remains difficult to express a specific definition. The World Business Council for Sustainable Development (WBCSD) defines CSR as the commitment of business to contribute to sustainable economic development working with employees, their families, the local community and society to improve their quality of life, in ways that are both good for business and good for development (Aramburu, 2019). Another definition made by the European Commission describes CSR as an integration of social and environmental concerns in their business operations and in the interrelation with the stakeholders on a voluntary basis (Rahman, 2011). Moreover, Kotler and Lee (2005) define CSR as a commitment to improve societal well-being through discretionary business practices and contributions of corporate resources. Mohr et al. (2001) indicate that CSR is the devotion of firms to maximize their positive effects or mitigate their negative effects to the society.

Regarding the approaches that researchers have used in order to conceptualize CSR, three main alternatives may be found (see Table 1). One of the most adopted approaches has been proposed by Carroll (1979, 1991) which states: *"the social responsibility of business encompasses the economic, legal, ethical, and discretionary (philanthropic) expectations that society has of organizations"*. Carroll (2016) conceptualized CSR as a pyramid where economic responsibility is the foundation of the pyramid, the second level is the legal obligation, the third level is the ethical responsibility, while at the top level is the philanthropic responsibility. Carroll (2016) also noted that economic and legal responsibilities are required, while ethical and philanthropic are expected. These responsibilities are mainly left to corporate judgment and choice, however, the expectation of business to achieve these goals is driven by social norms (Mandhachitara, 2011). Hence, the actions that an industry take, desiring to engage into social roles, are not dictated by law, but are driven by strategic orientation. Consequently, a firm that engages in CSR, should try to make profit, participate in ethical practices, be a good business citizen, and abide by law (Carroll, 2016).

Another approach structures CSR into three dimensions which include social (people), economic (profit), and environmental (planet) responsibilities (Beracs & Moisescu, 2015; Elkington, 1998). Elkington (1998) states that companies should focus on the environment which should be considered an organizational stakeholder besides creating profit or social values. Thus, a new dimension of CSR is introduced, i.e., environmental responsibility, where institutions are responsible for minimizing the impacts of business operations, protecting the environment, and managing the natural resources.

The third approach is the Stakeholder Theory (Freeman et al., 2010). This theory defines stakeholders as those groups or individuals that can affect or being affected by performance and activities of the organization (Pérez & Rodríguez, 2015a, 2015b). In this theory, CSR is described as the collection of initiatives that a firm engages in order to comply toward these groups instead of managing broader social concerns (Table 1).

It is understood that there are plenty of theories and approaches concerning CSR. Garriga and Mele (2004) classified the CSR theories into four groups: The first one is the instrumental theories where the firm

Table 1 Dimensional structure of CSR

Perspective	*Dimensions*
CSR pyramid	Economic responsibility Legal responsibility Ethical responsibility Philanthropic responsibility
Triple bottom-line approach	Social Economic Environment
Stakeholder theory	CSR for shareholders CSR for customers CSR for employees CSR for society General (legal, ethical)

is seen exclusively as an instrument for profit creation and the social activities are considered as means for economic growth. The second group consists of the political theories and concerns the power of the corporations in the society and the responsible use of this power into politics. The integrative theories are the third group in which the corporation focuses on satisfying social demands. Finally, the last group is the ethical theories which is based on the ethical responsibilities of an institution toward society.

3 Customer Satisfaction

Customer satisfaction has a crucial role in today's business management and is viewed as a valuable form of customer feedback (Ahmad et al. 2021; Wang, 2020). Oliver (2010) defines customer satisfaction as *"the consumer's fulfillment response. It is a judgment that a product or service feature, or the product or service itself, provided (or is providing) a pleasurable level of consumption-related fulfillment, including levels of under- or over-fulfillment"*. Furthermore, Oliver (1980) introduced the expectancy disconfirmation theory which has been widely used to study customer satisfaction. This theory proposes that customer satisfaction is the result of comparing a product or service's perceived performance with an individual's initial expectations (Wang, 2020).

Moreover, Gray and Boshoff (2004) identified that satisfaction lies in the perceptions of the customers toward the products or services. Hence, different customers may express different levels of satisfaction for the same product or services. Consequently, customer satisfaction is extremely important for firms in order to establish long-term relationships with customers (Gruca & Rego, 2005). Studies have indicated a positive correlation between strong long-term relationships with customers to larger profit and market value (Fornell et al., 2016).

Grigoroudis and Siskos (2010) present an overview of alternative customer satisfaction definitions and a detailed discussion about different customer satisfaction measurement approaches.

4 Customer Loyalty

Customer loyalty can be defined as: *"the strength of a customer's dispositional attachment to a brand (or service) and his/her intent to rebuy the brand (or re-patronize the service) in the future"* (Pan et al., 2012). Customer loyalty has a crucial role in firms, especially in periods of difficult economic situations (Pérez et al., 2013b). In this regard, customers are the most limited and valuable resource for companies, and thus loyalty is an important factor for profit, market growth, and competitive advantage (Kotler & Armstrong, 2010).

Three approaches may be distinguished for measuring customer loyalty: the behavioral, the attitudinal, and the composite approaches (Buttle, 2009; Rai, 2013). The behavioral approach suggests that the frequency of purchasing a product or service indicates customer loyalty toward the firm (Ehrenberg, 2000). The amount and the possibility of purchasing are the measures for the behavioral approach. Regarding the attitudinal approach, it deals with the emotion, beliefs, and preferences of customers, and therefore it has an emotional and a longer-lasting relationship (Buttle, 2009). The level of customer loyalty in this approach is measured by the willingness to pay a premium price, the possibility of repurchasing and the positive word of mouth. The composite approach combines the abovementioned approaches and measures loyalty through the repeat of purchase, the possibility of switching brand, and the product/service preference.

5 Corporate Social Responsibility, Customer Satisfaction, and Customer Loyalty in the Banking Sector

The results of the literature review regarding the role of CSR in customer satisfaction and customer loyalty are presented in this section.

Salmones et al. (2009) analyzed the influence of ethical and philanthropic responsibility of a financial entity on diverse constructs, such as customer satisfaction, trust, identification with the firm, business performance relational outcomes, and customer loyalty. They conducted personal surveys and obtained 789 valid responses, while the performed analysis was based on a structural equation modeling approach. The results indicated that loyalty toward a financial entity is directly determined by satisfaction, trust, and identification with the entity's values. Moreover, they found that there is an indirect link between CSR perceptions and customer satisfaction.

In a different context, Bravo et al. (2009) examined the impact of the corporate image of financial institutions on customer behavior by focusing on the differences between customers and non-customers of banking institutions. Collected data was based on a sample size of 450 individuals from five commercial banks in Spain. They considered CSR as one-dimensional element in their model which also includes elements such as global impression, location, services offered, accessibility, and personnel. The analysis showed that CSR has not a significant role toward satisfaction and intention to use a bank, but it can enhance the institution's overall image.

Poolthong and Mandhachitara (2009) investigated how socially responsible initiatives influence service quality and brand effect, as well as the role of trust as a mediating variable between perceived service quality and brand effect. 275 customers of retail banks in Bangkok, Thailand participated in this research. The analysis was performed by partial least squares (PLS) regression model. Moreover, they adopted three elements of Carroll's framework which deal with economic, ethical, and philanthropic dimensions. The results suggested that CSR positively affects customer attitudes toward a company and the quality of its service offerings. They highlighted that providing quality products and services, focusing on customer satisfaction and community support enhances the likelihood of positive perception of bank's service quality.

Mcdonald and Lai (2011) investigated the effects of three different types of CSR initiatives on Taiwanese retail banking customers' attitude and behavior. Three main categories of initiatives were included: customer-centric, philanthropic, and environmental initiatives. In order to collect the data, a questionnaire was designed, including several initiatives for each of the abovementioned categories. The total number of participants were 130 and the analysis of the results was made using the SPSS software. The main results indicated that the customers' attitude and behavior was highest for customer-centric initiatives, followed by philanthropic initiatives, while the environmental initiatives are being less preferred. More specifically, they found that customers preferred initiatives that benefit more themselves rather than those favoring other stakeholder groups. Finally, Mcdonald and Lai (2011) indicate that attitude is a component of customer satisfaction, hence, their findings hint that CSR activities improve customer satisfaction and loyalty.

Matute-Vallejo et al. (2011) explored the link between CSR activities and the price fairness to customer loyalty through satisfaction and commitment. To test their model, they employed structural equation modeling on a sample of 300 bank customers in Spain. The collection of data was carried through telephone questionnaires. They adopted the three-dimensional approach for CSR (social, economic, and environmental). The obtained results indicated that there is a positive connection between CSR and customer loyalty through satisfaction and commitment and this connection depended on the bank's orientation to its environment and society.

Senthikumar et al. (2011) explored the perception of customers on CSR in banking services in India. More specifically, they examined the relationship between CSR and customer satisfaction and their influence on service quality. The data sample was 1,200 customer, where 500 are customers of public sector bank, 400 are customers of private sector bank, and 300 are from cooperative banks. They developed a conceptual model of nine dimensions which includes CSR, customer satisfaction, and service quality and the analysis suggested that customer satisfaction is the most significant predictor of banking service quality, while CSR has a positive influence on customer satisfaction toward banking service quality.

The study of Mandhachitara and Poolthong (2011) examined the roles of CSR and perceived service quality in determining the attitudinal and behavioral loyalty of customers in the retail banking sector in Thailand. They analyzed the responses of 275 bank customers with the use of PLS regression. Furthermore, they incorporated three elements of Carroll's framework: economic, ethical, and philanthropic dimensions. The results demonstrated that CSR has a strong and positive association with attitudinal loyalty. Regarding behavioral loyalty, the perceived service quality mediated the relationship between CSR and behavioral loyalty.

Alafi and Hasoneh (2012) examined the relationship between CSR services and customer satisfaction with housing banks in Jordan. They used 18 items that define social responsibility and found that there is a positively significant relationship between CSR services and customer satisfaction. They also showed that there is a positive relationship among CSR and customer satisfaction to the financial performance of banks. The sample size was 203 and the SPSS software was used to analyze the data.

Pérez et al. (2013a) studied the relationship between corporate associations by examining the role of identification with the company and satisfaction in this connection. They used a structural equation model in a sample of 782 financial services users. Moreover, in order to measure corporate associations of CSR, they incorporated legal, ethical, and philanthropic elements from Carroll's (1979) model. The main outcomes of their research showed that commercial expertise is one of the most important determinants of both customer satisfaction and identification with their financial services provider. Furthermore, they found that CSR contributes to building customer identification with company, which is directly linked to satisfaction too. Finally, they concluded that satisfaction along with identification influences the attitudinal loyalty a customer shows toward their financial institution.

Pérez et al. (2013b) investigated the expectations of CSR during the crisis of the Spanish banking industry and the role of corporate governance in customer CSR expectations. They analyzed 684 responses of saving banks customers and 476 customers of commercial banks. Also, they adapted the stakeholder theory in order to measure CSR. The authors concluded that the customers of both types of banks have high expectations regarding CSR oriented to customers, shareholders, employees, community, and legal/ethical CSR. Moreover, customers of both types of banking companies are classified as customer-oriented, legally (customer) oriented, and CSR-oriented.

Polychronidou et al. (2014) empirically investigated customer's perception regarding CSR policies of Greek banks. The sample consisted of 113 customers who gave their opinion over different aspects of CSR, while collected data were analyzed using descriptive statistics and the SPSS software. The results indicated that most of the respondents expressed satisfaction for the CSR activities their bank is using, but they would not change their bank because of the CSR program.

Chomvilailuk and Butcher (2014) investigated the impact of CSR on three aspects of customer loyalty: word of mouth, purchase intention, and affective commitment for a new bank service. 204 bank customers in Australia participated in this survey. The CSR initiatives were described with statements attesting to the range of CSR beneficiaries, together with an evaluation on how long term was this new CSR initiative. The main results indicated that new CSR activities that come along with a new bank service had a positive impact on word of mouth and purchase intention, but it did not play an insignificant role in affective commitment. This research was the first one to examine loyalty using a multi-dimensional approach.

Another study that was held in Pakistan can be found in Khan et al. (2015). The purpose of their research was to analyze the impact of CSR perceptions on the perceived service quality and loyalty. They collected data from 480 customers from five different banks and tested their model by using PLS-based structural equation modeling. Furthermore, they considered three of the CSR construct that Carroll proposed (ethical CSR, legal CSR, and philanthropic CSR). Also, they measured customer loyalty through repurchase and word-of-mouth intensions. Their results showed that CSR is a direct determinant of perceived service quality, trust, repurchase and word-of-mouth intentions.

Pérez and Del Bosque (2015) investigated how the customer perceptions of the social responsibility of companies influence customer affective and conative responses through a hierarch of effects model. 1.124 customers of banking services in Spain participated in the survey. They incorporated the stakeholders theory and used the CSR structure proposed by Pérez et al. (2013). Their findings suggested that CSR directly influences customers' identification with the company and satisfaction, which lead to a positive effect to behavioral loyalty of customers.

Al-Ghamdi and Badawi (2019) examined the link between CSR and customer satisfaction and loyalty. They conducted their research to bank customers in Saudi Arabia (624 participants) by implementing a survey research strategy and analyzed the data with the SPSS software. The survey included questions regarding CSR activities that a bank is implementing, as well as questions that measure satisfaction and loyalty. The results suggested that CSR activities are enhancing both customer satisfaction and customer loyalty. Also, their research showed that there is a strong relationship between customer satisfaction and customer loyalty.

The study of Aramburu and Pescador (2019) examines the meditating role of corporate reputation on the relationship between CSR and customer loyalty. They took into consideration the role played by the bank type in the meditating effect. They collected 572 responses from bank customers in the Basque country. 118 customers evaluated commercial banks, 176 customers evaluated the Basque Credit Cooperative, and 278 customers evaluated saving banks. The surveys were completed using in-depth interviews. The CSR concept that was used in this research was composed of economic, social, and environmental items. The findings indicated that a bank's CSR behavior has greater influence on customer's attitudinal loyalty. Particularly, the social initiatives contribute more to loyalty compared to the other two dimensions. On the other hand, the bank type does not seem to affect the mediation effect.

Bugandwa et al. (2020) explored the association of CSR activities and trust. They operationalized CSR into five factors: legal responsibility, social responsibility, product responsibility, environmental responsibility, and employee responsibility. The sample size was 264 bank customers in the Democratic Republic of Congo (DRC), and the data were processed using exploratory and confirmatory factor analyses and structural equation modeling. Their research showed that each CSR dimension had a positive impact on customer's perception of trustworthiness.

Raza et al. (2020) examined the direct relationship between customer's perceptions of CSR and customer loyalty in the Pakistani banks. They collected 280 responses from three different banks in three large cities in Pakistan through field surveys. Moreover, the data analysis was executed with PLS-based structural equation modeling. The dimensions that measured CSR are: CSR related to customers, CSR related to shareholders, CSR related to employees, CSR related to society, and CSR related to ethical-legal issues. The findings showed that there is an insignificant relationship between CSR and customer loyalty. However, they found that electronic service quality, trust, and the customer-company identification play a meditating role in enhancing customer loyalty toward CSR activities.

Vo et al. (2020) investigated the impact of CSR on customer loyalty in the banking sector in Vietnam. They conducted onsite and online surveys regarding four commercial banks in Vietnam and collected 368 responses. For their data analysis, they used the SPSS software, applying exploratory factor analysis, confirmatory factor analysis, and structural equation modeling. They also considered the CSR approach of Carroll that includes four dimensions, and they added a customer-centric responsibility dimension. They concluded that the philanthropic responsibility, followed by the customer-centric responsibility have the most substantial impact on customer loyalty.

Zhang (2020) examined the effects of CSR on organizational performance by exploring the relationships between CSR, corporate reputation, customer satisfaction, and organizational attractiveness from the perspectives of both customers and job seekers. The data were collected through an online survey of 500 individuals that are employed in various firms, including banks. They used a three-dimensional approach for CSR, which includes CSR for employees, CSR for customers, and CSR for social public welfare. Their findings showed that CSR affects customer satisfaction, with a mediating role from corporate reputation.

A more recent study in the banking industry of Pakistan was conducted by Ahmad et al. (2021). They examined the effect of CSR on customer satisfaction and loyalty, as well as the mediating role of customer satisfaction and the moderating effect of corporate image. Moreover, they adapted the CSR scale with four items as proposed by Carroll and Shabana (2010). The data sample was 302 and the PLS-SEM modeling was used for their analysis. The main findings of their research suggest that CSR positively affects customer satisfaction, but it has an insignificant association with customer loyalty. They also indicated that customer satisfaction stimulates customer loyalty.

6 CSR and Mutual Funds

The effects of CSR practices in mutual funds are discussed in this section. The research in this field examines if and how investors derive utility from non-performance attributes of mutual funds.

In this context, Peifer (2014) in his study examined how economic and ethical concerns affect shareholder investment behavior using survey data from investors in socially responsible mutual funds. To a greater extent, they analyzed the levels of investor fund loyalty which is defined as the continued investment in a mutual fund, despite the belief that one has a lower return on investment. Their research indicates that investors are more loyal to their socially responsible funds than to their conventional funds. In addition, they demonstrated that economic motivation reduces socially responsible fund loyalty, while ethical motivation induces socially responsible fund loyalty.

Sandberg and Nilsson (2015) examined the ethical preferences of investors that engage to ethical or socially responsible investment profiled mutual funds. They developed a questionnaire that included a range of questions in order to determine participants' agreement with moral purity and moral effectiveness perspectives. They concluded that for financial services providers, there is a confusion in choosing between these dilemmas, even though they support both perspectives. As a result, financial providers are facing difficulties in which strategies and methods should incorporate in their ethical investment services.

In a more recent study, Li et al. (2021) examined the role of mutual funds in CSR. They evaluated 238 firms, 987 funds, and 921 CSR-related shareholder proposals. More specifically, each firm was evaluated on a set of CSR dimensions: community, diversity, employee relations, environment, human rights, and product safety. They found that CSR-friendly mutual funds enhance firms' CSR standings. Moreover, they discovered that CSR-friendly funds influence almost all the elements of CSR and focus on increasing CSR strengths. Also, they suggest that actively managed funds that were considered indifferent to social and ethical issues, play an important role in corporate social results of the firms they invest in.

7 Discussion

The presented literature review is based on a total of 24 papers, focusing either on the banking sector or mutual funds. The review covers a period from 2009 to 2021. Even though CSR as a concept has been radically developed over the last years, there is limited literature that deals with the immediate relationship between corporate social responsibility, customer satisfaction and customer loyalty. However, the examined papers are representative and cover this topic sufficiently. The examined studies applied different modeling approaches in order to examine the abovementioned relationships, while most of them examine the mediating role of other factors, such as corporate image, trust, identification, service quality, and customer-company initiatives (C–C).

An important factor that determines the relationship between CSR and customer satisfaction and customer loyalty is the perception of CSR. Most of the studies have adopted alternatives of Carroll's CSR pyramid or the stakeholder theory, but each study incorporated different aspects of CSR. Under this context, the dynamic nature of CSR and the difficulty of giving a "strict" description of CSR is confirmed. This might be explained by the fact that the existing literature expands into countries from different regions (developed and developing countries). Each country has different banking systems that are engaging into different activities. Nonetheless, the examined articles overall accept the multidimensional structure of the CSR concept. Table 2 summarizes the dimensions of CSR that each study took into consideration. It is observed that the dimensions with the highest frequency are the social, ethical, legal, and philanthropic responsibilities, while fewer studies have incorporated the environmental dimension.

Almost all of the examined papers did not incorporate together the concepts of customer satisfaction and customer loyalty (5 papers examined combined customer satisfaction and loyalty measures). In total, 10 papers investigated a direct link of CSR and customer satisfaction and 12 the link between CSR and customer loyalty. 2 out 10 did not support that CSR perceptions have a significant role on customer satisfaction. Regarding customer loyalty, only 3 studies did not find a direct link between CSR and customer loyalty or aspects of customer loyalty. At this point, it should be clarified that the majority of the studies examined on one hand the linkage between CSR and customer satisfaction or customer loyalty, but they also verified the mediating role of these concepts to service quality, trust, financial performance customer-company identification, and corporate reputation among others. Moreover, many of the studies confirmed a significant and positive association between customer satisfaction and customer loyalty.

The examined literature suggests that CSR initiatives of banking institutions have a positive effect on customer satisfaction and customer loyalty, regardless the CSR actions they invest in. It appears that CSR can be considered as an effective way of improving satisfaction and loyalty as customers are becoming more aware and pay more attention to societal obligations. In this way, bank institutions are able to build strong and long-term relationships with their customers, resulting to higher market value and enhanced organizational performance. Moreover, banks offer little space for product/service variation, and they rely on brand and

Table 2 Dimensions of CSR in the literature

Authors	*Perspective*	*Dimensions of CSR*
Salmones et al. (2009)	Based on Carroll CSR pyramid	Ethical responsibility Philanthropic responsibility
Bravo et al. (2009)	One-dimensional	-
Poolthong and Mandhachitara (2009)	Based on Carroll CSR pyramid	Economic responsibility Ethical responsibility Philanthropic responsibility
Mcdonald and Lai (2011)	Multidimensional	Customer centric initiatives Environment-protection initiatives Philanthropic initiatives
Matute-Vallejo et al. (2011)	Triple bottom line approach	Social Economic Environment
Senthikumar et al. (2011)	One-dimensional	-
Mandhachitara and Poolthong (2011)	Based on Carroll CSR pyramid	Economic responsibility Ethical responsibility Philanthropic responsibility
Alafi and Hasoneh (2012)	One-dimensional	Social responsibly initiatives
Pérez et al. (2013a)	Based on Carroll CSR pyramid	Legal responsibility Ethical responsibility Philanthropic responsibility
Pérez et al. (2013b)	Stakeholder theory	CSR for shareholders CSR for customers CSR for employees CSR for society General (legal, ethical)
Chomvilailuk and Butcher (2014)	Examined CSR perspectives through a questionnaire	-
Khan et al. (2015)	Based on Carroll CSR pyramid	Legal responsibility Ethical responsibility Philanthropic responsibility

(continued)

Table 2 (continued)

Authors	*Perspective*	*Dimensions of CSR*
Pérez and Del Bosque (2015)	Stakeholder theory	CSR for shareholders CSR for customers CSR for employees CSR for society General (legal, ethical)
Al-Ghamdi and Badawi (2019)	Multidimensional	Ethical activities Environmental activities Philanthropic activities
Aramburu & Pescador (2019)	Triple bottom line approach	Social Economic Environment
Raza et al. (2020)	Stakeholder theory	CSR for shareholders CSR for customers CSR for employees CSR for society General (legal, ethical)
Zhang (2020)	Based on Stakeholder theory	CSR for customers CSR for employees CSR for society
Bugandwa et al. (2020)	Multidimensional	Legal responsibility Social responsibility Product responsibility, Environmental responsibility Employee responsibility
Vo et al. (2020)	Based on Carroll CSR pyramid	Economic responsibility Legal responsibility Ethical responsibility Philanthropic responsibility Customer-centric responsibility
Ahmad et al. (2021)	Based Carroll CSR pyramid	Economic responsibility Legal responsibility Ethical responsibility Philanthropic responsibility

corporate image and therefore CSR is offered for creating differentiation. Table 3 summarizes the main relationships tested in literature and the major findings.

Table 3 Summary of the main relationships tested in the literature

Authors	*Relationships tested*	*Findings*
Salmones et al. (2009)	Relational outcomes → Satisfaction Commercial performance → Satisfaction Ethical resp. → Trust Philanthropic resp. → Identification	All the relationships are confirmed. Loyalty is determined through trust, satisfaction, and identification
Bravo et al. (2009)	CSR perceptions → C–C identification CSR perceptions → Satisfaction C–C identification → Loyalty Satisfaction → Loyalty	All the relationships are validated, except the relationship between CSR perceptions and customer satisfaction
Poolthong and Mandhachitara (2009)	CSR → Perceived Service Quality Perceived Service Quality → Trust CSR → Brand Affect Trust → Brand Affect	All the relationships are confirmed. Focusing on customer satisfaction and community support improves the positive perception of bank's service quality
Mcdonald and Lai (2011)	C–C initiatives will have stronger effect on customer attitude than philanthropic and environmental initiatives C–C initiatives will have stronger effect on customer behavioral intentions than philanthropic and environmental initiatives CSR initiatives impact behavior	All the relationships are confirmed
Matute-Vallejo et al. (2011)	CSR perceptions → Satisfaction Satisfaction → Loyalty	All the relationships are confirmed
Senthikumar et al. (2011)	CSR → Satisfaction Satisfaction → Service quality	All the relationships are confirmed
Mandhachitara and Poolthong (2011)	CSR perceptions → Repeat repurchase CSR perceptions → Attitudinal loyalty	CSR perceptions have a direct link to attitudinal loyalty, but do not impact repurchase intention
Alafi and Hasoneh (2012)	CSR → Customer Satisfaction → Financial Performance CSR → Customer Satisfaction Customer Satisfaction → Financial Performance CSR → Financial Performance	All the relationships are confirmed

(continued)

Table 3 (continued)

Authors	*Relationships tested*	*Findings*
Pérez et al. (2013a)	CSR perceptions → C–C identification CSR perceptions → Satisfaction C–C identification → Satisfaction C–C identification → Loyalty Satisfaction → Loyalty	All the relationships are confirmed, except the relationship between CSR perceptions and customer satisfaction
Chomvilailuk and Butcher (2014)	Perceptions of CSR performance → Loyalty Perceptions of new CSR info → Loyalty Perceived service quality → Loyalty	Loyalty is examined in three levels: purchase intentions, word of mouth, affective commitment. All the relationships are confirmed except the relationship between perceptions of CSR and affective commitment
Khan et al. (2015)	CSR perceptions → Perceived service quality CSR perceptions → Word of mouth intentions CSR perceptions → Trust CSR perceptions → Repurchase intentions	All the relationships are confirmed
Pérez and Del Bosque (2015)	CSR image → Satisfaction CSR image → C–C identification Satisfaction → Loyalty	All the relationships are confirmed
Al-ghamdi and Badawi (2019)	CSR activities → Customer satisfaction CSR activities → Customer loyalty Customer satisfaction → Customer loyalty	All the relationships are confirmed
Aramburu & Pescador (2019)	CSR → Customer loyalty CSR → Corporate Reputation Corporate reputation → customer loyalty	All the relationships are confirmed
Raza et al. (2020)	CSR → Electronic service quality CSR → Customer-company identification CSR → Trust CSR → Customer loyalty	All the relationships are confirmed, except the direct link of CSR and customer loyalty

(continued)

Table 3 (continued)

Authors	*Relationships tested*	*Findings*
Zhang (2020)	CSR → Customer satisfaction CSR → Organizational attractiveness CSR → Corporate reputation	All the relationships are confirmed
Bugandwa et al., (2020)	Product responsibility → Trust Legal responsibility → Trust Needs responsibility → Trust Environmental responsibility → Trust Employee responsibility → Trust	All the relationships are confirmed
Vo et al. (2020)	CSR → Corporate reputation CSR → Customer loyalty Corporate reputation → Customer loyalty	All the relationships are confirmed
Ahmad et al. (2021)	CSR → Customer loyalty CSR → Customer satisfaction Customer satisfaction → Customer loyalty CSR → Corporate reputation → Customer loyalty	All the relationships are confirmed except the mediating role of corporate reputation on customer loyalty

8 Conclusion and Future Research

The aim of this study is to examine the existing literature on the relationship between CSR, customer satisfaction, and customer loyalty in the banking sector. The findings of the research indicate the positive impact of CSR in both customer satisfaction and loyalty. These linkages function as an important factor for better financial performance and corporate reputation. Also, it is argued that by adopting and improving CSR activities, bank institutions can acquire a more satisfied and long-term customer base. Furthermore, this study can be used by future researchers and practitioners as a guide for developing new models for CSR activities and its implications in the banking sector.

The limited number of papers that focus on CSR and its impact on satisfaction and customer loyalty in the banking sector indicates that this is an open field of research. Future studies can verify the positive relationships found so far. Moreover, the studies that have been conducted so far do not include large sample sizes, therefore, future research may consider not only larger sample sizes, but also more representative samples. This will help to examine if customers are fully informed on their bank's services and CSR practices.

Finally, all of the studies have examined the various relationships using descriptive statistics and structured equations modeling. However, given the multidimensional nature of CSR, multicriteria decision analysis (MCDA) seems appropriate for analyzing and measuring CSR activities. In a similar way, MCDA can also be used to particular aspects of customer satisfaction. In this MCDA context, future studies may link specific CSR activities and customer satisfaction criteria.

References

Ahmad, M. J., Jawaid, A., Arshad, M. Z., & Paracha, S. H. (2021). Corporate social responsibility and customer loyalty: Exploring the role of satisfaction and corporate image in the banking industry. *Market Forces, 16*(1), 18–18. https://doi.org/10.51153/mf.v16i1.480

Al-ghamdi, S. A. A., & Badawi, N. S. (2019). Do corporate social responsibility activities enhance customer satisfaction and customer loyalty? Cogent Business & Management, 00(00). https://doi.org/10.1080/23311975.2019.1662932

Alafi, K., & Hasoneh, A. B. (2012). Corporate social responsibility associated with customer satisfaction and financial performance a case study with housing banks in Jordan. *International Journal of Humanities and Social Science, 2*(15), 102–115.

Aramburu, I. A. (2019). The effects of corporate social responsibility on customer loyalty: The mediating effect of reputation in cooperative banks versus commercial banks in the Basque Country. 701–719. https://doi.org/10.1007/s10551-017-3438-1

Aramburu, I. A., & Pescador, I. G. (2019). The effects of corporate social responsibility on customer loyalty: The mediating effect of reputation in cooperative banks versus commercial banks in the basque country. *Journal of Business Ethics, 154*, 701–719. https://doi.org/10.1007/s10551-017-3438-1

Beracs, J., & Moisescu, O. I. (2015). Feb.). Exploring the effect of perceived CSR on customer loyalty in the dairy products market. *Romanian Journal of Marketing, 2*, 28–35.

Bravo, R., Montaner, T. & Pina, J. M. (2009). The role of bank image for customers versus non-customers. *International Journal of Bank Marketing, 27*(4), 315–334. https://doi.org/10.1108/02652320910968377

Bugandwa, T. C., Kanyurhi, E. B., Akonkwa, B. M., & D., & Haguma Mushigo, B. (2020). Linking corporate social responsibility to trust in the banking sector: Exploring disaggregated relations. *International Journal of Bank Marketing, 39*(4), 592–617. https://doi.org/10.1108/IJBM-04-2020-0209

Buttle, F. (2009). *Customer relationship management concept and technology*. Elsevier Ltd.

Carroll, A. B. (1979). A three-dimensional conceptual model of corporate performance. *Academy of Management Review, 4*(4), 497–505.

Carroll, A. B. (1991). The pyramid of corporate social responsibility: Toward the moral management of organizational stakeholders. *Business Horizons, 34*(4), 39–48.

Carroll, A. B. (1999). Corporate social responsibility: Evolution of a definitional construct. *Business and Society, 38*(3), 268–295. https://doi.org/10.1177/000765039903800303

Carroll, A. B. (2016). Carroll' s pyramid of CSR: Taking another look. *International Journal of Corporate Social Responsibility, 1–8*, https://doi.org/10.1186/s40991-016-0004-6

Carroll, A. B., & Shabana, K. M. (2010). The business case for corporate social responsibility: A review of concepts, research and practice. *International Journal of Management Reviews, 12*(1), 85–105. https://doi.org/10.1111/j.1468-2370.2009.00275.x

Chambers, C. L., & Day, R. (2009). The banking sector and CSR: An unholy alliance. *Financial Regulation International, 12*(9), 13–20.

Chomvilailuk, R., & Butcher, K. (2014). Effects of quality and corporate social responsibility on loyalty. *Service Industries Journal, 34*(11), 938–954. https://doi.org/10.1080/02642069.2014.915952

Elkington, J. (1998). Partnership from cannibals with forks: The triple bottom line of 21st-century business. *Environmental Quality Management, 8*(1), 37–51. https://doi.org/10.1002/tqem.3310080106

Ehrenberg, A. S. (2000). Repeat buying-facts: Theory and applications. *Journal of Empirical Generalizations in Marketing Science, 5*(2), 392–770.

Flavia, C., & Torres, E. (2005). *The Influence of Corporate Image on Consumer Trust Internet Banking, 15*(4), 447–470. https://doi.org/10.1108/10662240510615191

Fornell, C., Morgeson, F. V., & Hult, G. T. M. (2016). Stock returns on customer satisfaction do beat the market: Gauging the effect of a marketing intangible. *Journal of Marketing, 80*(5), 92–107.

Freeman, R. E., Parmar, B. L., Harrison, J. S., Wicks, A. C., DeColle, S., & Purnell, L. (2010). Stakeholder theory: The state of the art. *Academy of Management Annals, 4*(1), 403–445. https://doi.org/10.1080/19416520.2010.495581

Gray, B., & Boshoff, C. (2004). The relationships between service quality, customer satisfaction and buying intentions in the private hospital industry. *South African Journal of Business Management, 35*(4), 27–37.

Gruca, T. S., & Rego, L. L. (2005). Customer satisfaction, cash flow, and shareholder value. *Journal of Marketing, 69*(3), 115–130.

Garriga, E., & Melé, D. (2004). Corporate social responsibility theories: Mapping the territory. *Journal of Business Ethics, 53*(1), 51–71. https://doi.org/10.1023/B:BUSI.0000039399.90587.34

Grigoroudis, E., & Siskos, Y. (2010). Customer satisfaction evaluation methods for measuring and implementing service quality. In Management Science (Vol. 139). https://doi.org/10.1007/978-1-4419-1640-2

Kotler, P., & Lee, N. (2005). *Corporate social responsibility: Doing the most good for your company and your cause*. Wiley.

Kotler, P., & Armstrong, G. (2010). Principles of marketing: Pearson education.

Khan, Z., Ferguson, D., & Pérez, A. (2015). *Customer responses to CSR in the Pakistani banking industry, 33*(4), 471–493. https://doi.org/10.1108/IJBM-07-2014-0097

Li, F., & Z., Saurin, L., Patel S., and Ramani Srikanth,. (2021). The role of mutual funds in corporate social responsibility. *Journal of Business Ethics, 174*(3), 715–737. https://doi.org/10.1007/s10551-020-04618-x

Mandhachitara, R. (2011). A model of customer loyalty and corporate social responsibility. 2(August 2009), 122–133. https://doi.org/10.1108/08876041111119840

Mandhachitara, R., & Poolthong, Y. (2011). A model of customer loyalty and corporate social responsibility. *Journal of Services Marketing, 25*(2), 122–133. https://doi.org/10.1108/08876041111119840

Matute-Vallejo, J., Bravo, R., & Pina, J. M. (2011). The influence of corporate social responsibility and price fairness on customer behaviour: Evidence from the financial sector. *Corporate Social Responsibility and Environmental Management, 18*(6), 317–331. https://doi.org/10.1002/csr.247

Mcdonald, L. M. (2007). *Corporate social responsibility and bank customer satisfaction a research agenda*. https://doi.org/10.1108/02652320810864643

McDonald, L. M., & Hung Lai, C. (2011). Impact of corporate social responsibility initiatives on Taiwanese banking customers. *International Journal of Bank Marketing, 29*(1), 50–63. https://doi.org/10.1108/02652321111101374

Mohr, L. A., Webb, D. J., & Harris, K. E. (2001). Do consumers expect companies to be socially responsible? *The Impact of Corporate Social Responsibility on Buying Behavior, 35*(1), 45–72.

Oliver, R. L. (1980). A cognitive model of the antecedents and consequences of satisfaction decisions. *Journal of Marketing Research, 17*(4), 460–469.

Oliver, R. L. (2010). Customer satisfaction. *Wiley International Encyclopedia of Marketing*. https://doi.org/10.1002/9781444316568

Pan, Y., Sheng, S., & Xie, F. T. (2012). Antecedents of customer loyalty: An empirical synthesis and reexamination. *Journal of Retailing and Consumer Services, 19*(1), 150–158. https://doi.org/10.1016/j.jretconser.2011.11.004

Park, E., & Kim, K. J. (2019). What drives "customer loyalty" ? The role of corporate social responsibility. February 2018, 304–311. https://doi.org/10.1002/sd.1901

Peifer, J. L. (2014). Fund loyalty among socially responsible investors: The importance of the economic and ethical domains. 635–649. https://doi.org/10.1007/s10551-013-1746-7

Pérez, A., García de los Salmones, M. del M., & Rodríguez del Bosque, I. (2013a). The development of a stakeholder-based scale for measuring corporate social responsibility in the banking industry. 459–481. https://doi.org/10.1007/s11628-012-0171-9

Pérez, A., & García de los Salmones, M. del M., & Rodríguez del Bosque, I. (2013b). The effect of corporate associations on consumer behaviour. *European Journal of Marketing, 47*(1), 218–238. https://doi.org/10.1108/03090561311285529

Pérez, A., & Del Bosque, I. R. (2015). Customer responses to the CSR of banking companies. *Journal of Product and Brand Management, 24*(5), 481–493. https://doi.org/10.1108/JPBM-12-2014-0759

Pérez, A., & Rodríguez del Bosque, I. (2015a). How customer support for corporate social responsibility influences the image of companies: Evidence from the banking industry. 168(July 2013), 155–168. https://doi.org/10.1002/csr.1331

Pérez, A., & Rodríguez del Bosque, I. (2015b). An integrative framework to understand how CSR affects customer loyalty through identification. *Emotions and Satisfaction, 571–584*. https://doi.org/10.1007/s10551-014-2177-9

Platonova, E., Asutay, M., Dixon, R., & Mohammad, S. (2018). The impact of corporate social responsibility disclosure on financial performance: Evidence from the GCC Islamic banking sector. *Journal of Business Ethics, 151*(2), 451–471. https://doi.org/10.1007/s10551-016-3229-0

Polychronidou, P., Ioannidou, E., Kipouros, A., Tsourgiannis, L., & Simet, G. F. (2014). Corporate social responsibility in Greek banking sector—an empirical research. *Procedia Economics and Finance, 9*(Ebeec 2013), 193–199. https://doi.org/10.1016/s2212-5671(14)00020-3

Poolthong, Y. (2009). Customer expectations of CSR, perceived service quality and brand effect in Thai retail banking. 2006. https://doi.org/10.1108/02652320910988302

Poolthong, Y., & Mandhachitara, R. (2009). Customer expectations of CSR, perceived service quality and brand effect in Thai retail banking. *International*

Journal of Bank Marketing, 27(6), 408–427. https://doi.org/10.1108/02652320910988302

Rai, A. K. (2013). Customer relationship management: Concepts and cases (Vol. 2). PHI Learning.

Rahman, S. (2011). Evaluation of definitions: Ten dimensions of corporate social responsibility. *World Review of Business Research, 1*(1), 166–176.

Raza, A., Rather, R. A., Iqbal, M. K., & Bhutta, U. S. (2020). An assessment of corporate social responsibility on customer company identification and loyalty in banking industry: A PLS-SEM analysis. *Management Research Review, 43*(11), 1337–1370. https://doi.org/10.1108/MRR-08-2019-0341

Sandberg, J., & Nilsson, J. (2015). Do ethical investors want purity or effectiveness? An exploratory study on the ethical preferences of mutual fund investors. *Journal of Financial Services Marketing, 20*(1), 34–45. https://doi.org/10.1057/fsm.2015.3

Salmones, M., Perez, A., & del Bosque, I. (2009). *The social role of financial companies as a determinant of consumer behaviour, 27*(6), 467–485. https://doi.org/10.1108/02652320910988339

Senthikumar, N., Ananth, A., & Arulraj, A. (2011). *Impact of corporate social responsibility on customer satisfaction in banking service, 5*(April), 3028–3039. https://doi.org/10.5897/AJBM10.1678

Vo, H. D., Van, T. H. L., Dinh, T. H. L., & Ho, M. C. (2020). Financial inclusion, corporate social responsibility, and customer loyalty in the banking sector in Vietnam. *Journal of International Studies, 13*(4), 9–23. https://doi.org/10.14254/2071-8330.2020/13-4/1

Wang, C. C. (2020). Corporate social responsibility on customer behaviour: The mediating role of corporate image and customer satisfaction. *Total Quality Management and Business Excellence, 31*(7–8), 742–760. https://doi.org/10.1080/14783363.2018.1444985

Weber, M. (2008). The business case for corporate social responsibility: A company-level measurement approach for CSR Manuela Weber. 247–261. https://doi.org/10.1016/j.emj.2008.01.006

Zhang, Q. (2020). Effects of corporate social responsibility on customer satisfaction and organizational attractiveness: A signaling perspective. July 2019, 20–34. https://doi.org/10.1111/beer.12243

Socially and Environmentally Responsible Investments and Mutual Funds

Michalis Doumpos, Marianna Eskantar, and Constantin Zopounidis

1 Introduction

Corporate social responsibility (Friedman, 2007) has recently reappeared in the context of the dramatic increase in socially responsible investment (SRI). Assets managed in socially responsible capital have multiplied, and many "traditional" investors are considering increasing their asset allocation with Environmental, Social, and Governance Scores (ESG, Pastor et al., 2019). From the point of view of asset management, this trend

M. Doumpos (✉) · M. Eskantar · C. Zopounidis
School of Production Engineering and Management, Technical University of Crete, Chania, Greece
e-mail: mdoumpos@tuc.gr

M. Eskantar
e-mail: meskantar@tuc.gr

C. Zopounidis
e-mail: kzopounidis@tuc.gr

C. Zopounidis
Audencia Business School, Nantes, France

C. Gaganis et al. (eds.), *Sustainable Finance and ESG*, Palgrave Macmillan Studies in Banking and Financial Institutions,
https://doi.org/10.1007/978-3-031-24283-0_4

raises questions about the financial returns of such investments (Hong & Kacperczyk, 2009). However, if SRIs are to have a real impact, they must influence investment decisions and the operation of firms.

The concept of SRI seeks to introduce non-financial elements into investment decision-making. These elements address issues of ethics as well as social, environmental, and corporate governance. This type of investment has attracted high interest from academics, professionals, investors, and financial institutions such as banks. The amount of capital invested in funds that are intended to incorporate ethical concerns or concerns related to ESG criteria has increased in recent years. Over the years, it has become more and more understandable and clear that non-financial issues are of greater interest. However, there is a great deal of ambiguity as to exactly how specific features should be formulated and incorporated in the investment processes. In addition, there is now consensus on how SRI investments should be defined.

Some academics have criticized the idea of SRI for all the turmoil that have been created over the years, as well as for the integration of non-financial investment concerns because there is a lack of clarity in both the SRI objectives and the vagueness in the SRI framework. Investment professionals argue that SRI can be interpreted differently in each market and even in a specific market, different interpretations are often used. Therefore, it is crucial to have acceptable standards that define an SRI (Sandberg et al., 2009).

The purpose of this article is to provide an overview of SRI and its relationship to ESG. Moreover, we present a framework of criteria and indicators for defining and evaluating SRIs and overview the literature on SRI/ESG mutual fund investments, covering various issues, such as the characteristics of such investments, behavioral issues, performance assessment, and portfolio optimization.

The rest of the chapter is organized as follows. Section 2 provides a short discussion of the concept of SRI, followed in Sect. 3 by a historical background and how the relationship with ESG is directly linked. Section 4 presents a collection of ESG criteria that can provide evaluations to stakeholders, whereas Sect. 5 provides an overview of the literature on SRI/ESG investments in the mutual funds' industry. Finally, Sect. 6 concludes the chapter and discusses some future research directions.

2 Socially Responsible Investing

Most definitions of SRI relate it to "integrating personal values and social concerns with investment decisions" (Schueth, 2003). In addition to financial aspects, this definition includes the report of social evaluations for the selection of investment projects. The roots of SRI are traced back to the early nineteenth century in various religious movements, although the term was not well-known until the 1980s.

According to the Report on US Sustainable and Impact Investing Trends, the largest sustainable investment development had taken place since 2012 using ESG criteria and by 2020, the total assets exceeded $18,000 million. In terms of assets, money managers have incorporated social factors a little more than environmental and governance criteria. Social criteria incorporated by money managers have increased by 49% from 2018 to $16.1 trillion, whereas the consideration of environmental factors increased at a faster pace during 2018–2020, rising 57%, from $10.1 trillion to nearly $16.0 trillion.

Despite the rapid developments in SRIs, there is no clear definition of what exactly constitutes an SRI. Typical investors in SRI prefer to reward companies that exhibit positive social behavior. Religious values are also considered as important factors in SRI by investors. The foreclosure approach, which is the most popular, considers the products and specific corporate tactics, as filters in the portfolio selection process (Berry & Junkus, 2013). Common products that are excluded are alcohol, tobacco, gambling, and weapons. Cases of violation of labor legislation (e.g., child labor) are also employed as negative filters. On the other hand, the inclusion approach requires the adjustment of an investment depending on the behavior of the firm. In this approach, a consistency rating should be defined to act positively in the sense of SRI. This process has a downside as there is high subjectivity. Moreover, the investor should specify the rating criteria and the corresponding corporate behaviors, as well as their relative importance. Then, the performance of the companies on the selected criteria should be measured and the derived ratings should be related to the composition of the investment portfolio. Such difficulties pose challenges on the adoption of the inclusion approach, thus leading to the popularity of the simpler foreclosure approach. Nevertheless, given the various definitions of SRI that have been proposed over the years, instead of focusing on a specific definition, it makes sense to monitor how the SRI concept involves over time (Renneboog et al., 2008).

3 From SRI to ESG

The SRI concept refers to a set of values that originated from religious dogmas and evolved into a modern field that addresses issues of social justice, climate change, and environmental awareness, as well as corporate governance concerns.

Some movements that took place in the distant past that are reminiscent of the concept of social responsibility were originally the Quakers in the seventeenth century who refused to take advantage of the arms and slave trade when they settled in North America. Then John Wesley (1703–1791) stated that people should not engage in sinful trade or profit from the exploitation of others. As in the 1920s, the Methodist Church in the United Kingdom avoided investing companies that engaged in activities such as the production of alcohol, tobacco, weapons, and in gambling. However, the first modern mutual fund, founded in 1928, used religious traditions and was called the Pioneer Fund. The moral investment or responsible investment also seems to have originated from the Islamic tradition. Based on interpretations of the Koran, Muslim investors avoid investing in companies involved in pork production, pornography, gambling, and interest-based financial institutions.

The idea of corporate social responsibility (CSR), as well as the reference to it, was almost non-existent in 1968 when Moskowitz began his career, as the movements based on this idea were minimal and almost random. When Moskowitz began to address this issue, there were few sources of information immediately available. But Moskowitz succeeded and built a solid foundation for CSR on which decades of research could be based.

The traditional concept of SRI seems to have been heavily influenced by the 1960s and 1970s, as they saw the rise of the anti-war movement and the maturity of movements for racial equality, women's rights, consumer protection and the environment. In the late 1960s, the Vietnam War created a complication for investors, let alone the "responsible" investors of the time. By the 1970s, some in North America began looking for ways to avoid "war speculation" in their portfolios.

In the early 1970s, this led to the creation of the first mutual funds that reflected beliefs based on values, sensitivities of the time that seemed to relate to political rights as well as environmental concerns. The first modern SRI fund, called the Pax World Fund, was founded in 1971 in the United States. An indicative example of the environmental concerns that started raising during that period is the initiative taken by Wisconsin Sen, Gay Lord Nelson, and Dennis Hayes to manage to mobilize 20 million Americans on April 22, 1970, to celebrate Earth Day for the first time. That year, the Environmental Protection Service was created and the law on clean air was passed. After all this, laws were created concerning the environment and consumer protection. North American SRI was born in this context (Townsend, 2020).

In 1972, the $25 million Dreyfus Third Century fund, began looking for companies that carried out activities that, compared to other companies in the same industry, helped to improve the quality of life (Levering & Moskowitz, 1998). The fund's strategy relied on ranking the companies, thus leading to what is now known as the "best in class" analysis.

In the early 1990s, corporate social investment grew exponentially in the United States, Europe, and other parts of the world. A key factor that helped this development was moral consumerism. As an indicative example, it is worth noting that ethical consumer spending and finance in the United Kingdom, was estimated to be more than £120 billion in 2020.[1] Moreover, new factors are being introduced into the SRI concept as a series of corporate scandals have demonstrated the importance of new issues, such as transparency and governance as key dimensions for evaluating an SRI.

Modern SRI is divided into two categories. The first category involves value-based investments, i.e., those that focus on meeting the values of investors. The second category focuses on the analysis of the ESG pillars, to assess the importance of non-financial data for the selection and evaluation of SRIs. According to the 2020 Trend Report by the US Social Investment Forum (USSIF), US investments that consider ESG factor grew to $17.1 trillion at the start of 2020, compared to $12 trillion two years before, thus corresponding to 33% of the total US assets under management.[2] It is the rise of ESG investment that makes modern SRI more than just a market term of social responsibility. ESG is what traditional SRI cannot be. These new pillars help to better position the term SRI by evading only social issues and evolving the term into issues that directly concern companies, investors, banks, and the wider society.

[1] https://coop.uk/3S7TjuZ (last accessed: July 29, 2022).

[2] https://www.ussif.org/currentandpast (last accessed: July 29, 2022).

4 ESG Criteria for SRI

Given the multifaced and complex nature of ESG factors, in this section we compile a comprehensive list of indicators and criteria that can serve as a methodological basis for assessing the ESG performance of firms and selecting SRIs. The selection of the indicators presented below has been based on the existing literature on this area.

Environment

The environment pillar covers all aspects of a company's footprint on the environment. In this pillar, we consider three main categories of indicators, namely emissions, resource use, and environmental innovation. The emissions dimension concerns the willingness and ability of a company to implement effective actions for reducing its emissions. The resource dimension evaluates a company's performance on the reduction of materials, energy, and water usage. Finally, the third dimension of the environment pillar, refers to the capacity of a company to develop innovative environmental technologies, processes, and products.

(a) Emissions:

- Waste recycling.
- CO2 equivalent emissions, covering gases such as: carbon dioxide, methane, nitrous oxide, hydrofluorocarbons, etc.
- Total waste to revenues.
- Existence of environmental management systems (e.g., ISO 14000).
- Environmental fines concerning the violation of environmental regulations.
- Biodiversity impact and activities to reduce the impact of the company's activities on ecosystems and species.

(b) Use of resources:

- Water use to revenues.
- Water recycled (i.e., the amount of water recycled or reused).
- Land use, i.e., initiatives that a company takes to reduce its environmental impact on land owned, leased, or managed for production activities.

- Total energy use to revenues.
- Renewable energy use ratio.
- Implementation of policies regarding: (i) the impact of the supply chain on the environment, (ii) the reduction of toxic chemicals or substances, and (iii) sustainable packaging.

(c) Innovation:

- Environmental R&D expenditures.

Society

The society pillar involves issues related to human rights and the well-being of communities within which a company operates. Four categories are considered in this dimension. The workforce category describes a company's employment and job policies (e.g., safety, equal opportunities, development opportunities, etc.). The second category relates to human rights, i.e., policies regarding fundamental human rights conventions. The next category is community; it involves a company's commitment toward protecting public health, respecting business ethics, etc. The last category is product responsibility and reflects a company's capacity to produce quality goods and services, adopting health, safety, integrity, and data privacy principles.

(a) Workforce:

- Policies for employee health and safety.
- Initiatives taken to measure, monitor, and improve employees' health and safety in the supply chain (e.g., whether the company cooperates with suppliers that do not follow proper security standards for their employees).
- Policies regarding the training and career development of its employees.
- Policies toward ensuring diversity and equal opportunity on special groups of employees, such as women, minorities, disabled employees, etc.
- Employee satisfaction.
- Women employees and managers.
- Employees with disabilities.

(b) Human rights:

- Implementation of policies to ensure the respect of human rights in general.

(c) Community:

- Total donations to revenues.

(d) Product responsibility:

- Implementation of policies and processes to protect customer and public privacy and integrity.
- Responsible marketing ensuring the protection of children.
- Customer satisfaction.
- Products and services with specific health and safety benefits for the consumers.
- Existence of industry-specific quality certifications (e.g., ISO certifications).
- Product responsibility monitoring, i.e., whether a company monitors the impact of its products or services on consumers and the community.

Corporate Governance

The last pillar is corporate governance, which is analyzed in three sub-categories. The first category (management) focuses on the adherence of a company to the principles of corporate governance and its effectiveness in implementing relevant policies. The second category examines issues related to the participation of the shareholders in the administration of a company. The CSR strategy of a company is reflected on the indicators in the third group.

(a) Management:

- Board structure (e.g., unitary structure, two-tier structure, mixed two-tiered structure, etc.) and independence.
- Policy board experience.
- Existence of an audit board committee.
- Participation of non-executive and independent members to the audit committee.

(b) Shareholders:

- Existence of policies to facilitate shareholder engagement.
- Confidential voting policy.
- Election of board members and director through a majority voting process.

(c) CSR strategy:

- Existence of an external CSR auditor.
- Explicit integration of financial and extra-financial factors in the management discussion and analysis section of the company's annual report.

5 Literature Review on Social and Environmental Mutual Funds

The growing importance of SRIs in the financial markets, has led to a rich academic literature on this area. Focusing on the case of fund management, in the following we organize the existing literature into four broad streams. The first one focuses on the characteristics of mutual funds that adopt the principles of SRI and ESG. The second examines behavioral issues regarding the attitude of the investors toward SRI/ESG. The third group involves studies that examine the performance of SRI/ESG funds, whereas the final one focuses on the construction of SRI portfolios. This section provides a brief overview of each of these streams in the literature. More comprehensive reviews can be found in the works of Junkus and Berry (2015), Popescu et al. (2021), and Widyawati (2020).

Characteristics of Socially Responsible Mutual Funds

This first stream of the literature examines how and at what level mutual funds integrate the principles of SRI and ESG. The findings from the literature indicate that socially responsible (SR) funds do not always achieve their role as promoters of responsibility principles and there is still room for improvement on their true SR performance.

On the positive side, one can refer to studies such as those of van Duuren et al. (2016) and Dorfleitner et al. (2021). van Duuren et al. (2016) conducted an international survey among conventional fund managers to examine whether they consider ESG factors in their investment decisions. Their results shown that the principles of SRI/ESG are used by conventional funds as screening tools, similarly to fundamental investing. Nevertheless, they noticed that US managers were more skeptical about the financial materiality of SRI compared to managers in the UK and the EU.

Dorfleitner et al. (2021) used a data set of more than 400 SR mutual funds in the USA over the period 2003–2018, to examine their ESG and controversy performance over time. The latter refers to business practices that are considered as socially irresponsible or unethical (e.g., environmental scandals, business ethics controversies, etc.). The authors found that SR mutual funds show persistence in their ESG and controversy performance over time. However, they observed that there is a trade-off between ESG ratings and controversy scores, in the sense that funds with high ESG scores tend to perform poorly on the controversy dimension.

Regarding studies that report negative results on the actual SR character of SR mutual funds, one can cite studies such as those of Candelon et al. (2021), Raghunandan and Rajgopal (2022), and Utz and Wimmer (2014). Candelon et al. (2021) used a sample of 1500 SR mutual funds in the USA and the EU, to examine whether their investment practices actually comply with their stated ESG commitments. They found that fund managers often act in an opportunistic manner and despite claiming to act in accordance with SR principles, their investment strategies diverge from SRI/ESG standards. The authors refer to this phenomenon as "ESG-washing".

Raghunandan and Rajgopal (2022) used a sample of ESG mutual funds in the USA from 2010 to 2018 to examine whether they invest in firms that have stakeholder-friendly track records. They found that even though these funds have higher ESG scores compared to non-ESG funds, they invest in firms with worse track records for compliance with labor and environmental laws, compared to non-ESG funds. Similar results were also reported by Utz and Wimmer (2014) who found that, on average, SR mutual funds are not holding more ethical assets, and they do not ensure the exclusion of unethical firms. While the aforementioned studies rely on statistical descriptive models (e.g., regression analysis), Utz et al. (2014, 2015) followed a different path adopting a portfolio optimization approach that extends the standard mean–variance model with a third ESG criterion. The application of the optimization models on SR funds, showed that there is considerable room for improving the portfolio allocation process to increase the ESG performance of SR portfolios, compared to the current practices used by SR funds.

Behavioral Issues

A second part of the literature explores the attitudes of investors toward SRI funds, i.e., why investors choose sustainable investments and what characteristics they value most.

To explore these issues, Riedl and Smeets (2017) used data from a mutual fund provider in the USA as well as survey data from individual investors. They found that social preferences and social signaling are important factors for SRI decisions. On the other hand, financial issues were found to be of lower importance, as investors with strong social interest are willing to accept lower financial results for their investments.

Hawn et al. (2018) used an international sample covering 27 countries over the period 1999–2015 and found that investors have a limited response to news about events about the Dow Jones Sustainability Index (DJSI) World index (i.e., where a firm is added, deleted, or retained on the index).

Lapanan (2018) examined the trading behavior of investors in SR mutual funds. The analysis was based on a sample of individual investors in Sweden covering the period 2003–2007. According to the obtained results, SR investors hold more diversified portfolios that combine SR and conventional mutual funds. Moreover, portfolio rebalancing is done more often for SR investors than conventional ones. Moreover, the analysis showed that, compared to conventional investors, SR investors are less likely to sell their SR investment when past returns decrease, which indicates that SR investors value sustainability. Similar fundings were also reported by Bollen (2007), for a sample of US funds over the period 1980–2002. Matallín-Sáez et al. (2022) further examined this issue (reaching to similar conclusions as the previous studies), and linked it to the disposition effect, according to which investors are more inclined to sell their profitable investments as opposed to selling those with losses (Shefrin & Statman, 1985).

The Performance of SR Mutual Funds

Two sub-categories can be identified in this stream of literature on the performance of SR mutual funds. The first involves empirical descriptive studies examining the performance of SR funds, whereas the second is more methodologically oriented, focusing on approaches for performance assessment.

Regarding the empirical literature on the performance of SR funds, Climent and Soriano (2011) compared the SR, green, and conventional funds in the USA in terms of their performance and risk sensitivities, over the period 1987–2009. Based on a CAPM approach, the results showed that over the whole period, green funds had lower performance compared to conventional ones. The authors explained this result based on the smaller investment pools that SR and environmental funds use, which leads to higher risks. However, when focusing on the more recent period 2001–2009, no significant differences were observed between the different types of funds.

Utz and Wimmer (2014) used a large sample consisting of over 37,000 conventional funds and 230 SR funds over the period 2002–2012. Using standard financial performance measures, such as the return, the variance, the Sharpe and Treynor ratios, beta, as well as single and multi-index models, they found no significant differences between the performances of SR funds compared to their conventional peers.

While the previous studies focused on US funds, Cortez et al. (2012) used an international sample involving both US and European funds (46 funds overall), covering the period 1996–2008. The analysis was based on single and multi-index models to investigate the funds' returns in comparison to conventional and SR benchmark portfolios. According to the results, the European funds were not found to have significantly different performance compared to the benchmarks, whereas US funds performed worse. Similar results were also reported by Muñoz et al. (2014), who further examined the role of market conditions (i.e., normal conditions versus crisis) as well as the stock-picking and market-timing abilities of the funds' managers. The evidence provided in the study, showed that, in general, these abilities are missing from SR funds. However, some discrepancies from this general finding were observed depending on the region (USA, Europe) and the market conditions.

The performance of SR funds under crises conditions was more extensively analyzed by Nofsinger and Varma (2014), who examined this issue using data involving US funds for the period 2000–2011. During that period, the authors focused on two crises: the technology bubble burst of 2000–2002 and the global crisis of 2007–2009. The results of the study showed, that although during the non-crises periods, SR funds performed worse than conventional ones, they achieved superior performance during the crises. Further analysis showed that these results are mainly driven by the funds SR characteristics and the use of positive screening techniques, rather than the management of the portfolios. The effectiveness of the screening mechanisms employed by SR funds, especially during adverse market conditions, was also confirmed in a latter study by Henke (2016) for a sample of SR bond funds in the USA and the Eurozone. Such results are in accordance with those reported by Chen and Scholtens (2018), who found no performance benefit for actively managed SR funds over those that employ passive investment strategies. It is worth noting, however, that according to Geczy et al. (2021), the effect of introducing SR in fund management, is stronger when multi-index models are used for performance evaluation, which are often employed in the empirical finance literature.

Studies such as the ones mentioned above originate from the finance literature and adopt an empirical descriptive perspective using statistical and econometric approaches. A second stream of the literature on the performance of SR mutual funds focuses on the development and application of methodologies for assessing the performance of funds. Such approaches originate from the field of operations research/management science and typically employ methodologies based on multicriteria decision analysis (MCDA) and data envelopment analysis (DEA). Based on this research approach, García-Melón et al. (2016) proposed a MCDA framework for assessing the CSR performance of mutual funds from a multi-stakeholder perspective. The analytic hierarchy process was employed to define the weights of the evaluation criteria based on the information gathered through a series of interviews with various stakeholders. The methodology was used to evaluate the CSR performance of 37 mutual funds in Spain. Pérez-Gladish et al. (2013) used a DEA model to assess the efficiency of equity mutual funds combining financial and SRI criteria. The model was applied to conventional and SR funds with the results indicating no significant differences between the two types of funds. DEA-based models were also used in the studies of Basso and Funari (2014) and Allevi et al. (2019). Such methodologies can be used to design improved systems for rating the ESG/SRI performance of mutual funds. It is worth noting that currently available ratings from major agencies show considerable discrepancies (Gangi et al., 2022) due to the lack of commonly acceptable framework and procedure for such ratings.

Portfolio Optimization for SR Mutual Funds

The last stream of the literature that we consider in this review involve portfolio optimization issues in SRI. Traditional portfolio optimization approaches, such as the mean–variance model, rely on risk-return criteria. In the context of SRI/ESG, however, additional objectives should be considered to describe the SRI and ESG characteristics of the portfolios. Naturally, this leads to an extension of the bi-objective optimization formulations of the standard framework, to multi-objective problems.

Steuer et al. (2007) analyzed the theoretical issues and challenges that arise in the adaptation of the standard models to the multi-objective SRI/ESG setting. Later studies presented various models and formulations in this context. For instance, Ballestero et al. (2012) presented a goal programming (GP) approach, which allows investors to specify their investment preferences by setting targets on goals to be reached. Bilbao-Terol et al. (2012a, 2012b) also employed a GP model based on conditional value-at-risk, combined with fuzzy MCDA approach, which allows the modeling of fuzzy preferences on SRI goals.

In the above studies, the proposed optimization models were mainly used to illustrate the applicability of the aforementioned approaches in a prescriptive context. Nevertheless, portfolio optimization approaches in SRI/ESG have also been employed in a descriptive and predictive framework. For instance, building on the framework presented by Steuer et al. (2007), the studies of Utz et al. (2014, 2015) presented portfolio optimization models based on three objectives, namely return, variance, ESG performance. Except for the theoretical development of the three-objective models, the authors applied them to samples of SRI funds from the USA, finding that the current practices of the funds on the ESG dimension is much lower than what can be achieved if proper portfolio optimization approaches are adopted. Gasser et al. (2017) also considered a three-objective portfolio optimization approach and through its application to a large sample of more than 6,000 companies, they found that there is indeed a trade-off between financial and SR objectives. This trade-off cannot be accommodated through traditional risk-return optimization, which leads to significantly lower SR ratings.

Finally, with exploitation of big data and the widespread use of modern computational technologies from the area of artificial intelligence, new possibilities arise in developing sophisticated portfolio optimization approaches for SR funds. Toward this direction, Vo et al.

(2019) presented an optimization architecture for SR investments which combines three components: (i) a deep learning system (a bi-direction long short-term memory neural network) to forecast stock returns, (ii) a mean–variance-ESG portfolio optimization model for capital allocation, and (iii) a reinforcement learning module to retrain the prediction system and rebalance the portfolios in a dynamic context.

6 Concluding Remarks

SRI is a process that considers the social, environmental, and corporate implications of financial investments (Olmedo et al., 2010). Despite the development of SRI, there is still a strong need for reliable information the environmental policies that companies follow, the social impact, and their governance strategies. Having access to such information is fundamental for attracting investors in SRIs.

The need for ESG information has led to the emergence of a plethora of data providers and agencies that provide information to investors on various sustainability indicators. These sustainability indicators help to study the economic, social, environmental, and corporate governance performance of companies. For these indicators to be helpful in practice, it is important that they provide an objective view of the principles of CSR, SRI, and ESG through measurable metrics. Based on this idea, in this chapter we provided a comprehensive list of evaluation criteria that can be used for this purpose. Such a list can serve as a basis for developing comprehensive evaluation systems.

The area of SRIs and ESG is rapidly progressing as the global industry faces new challenges. For instance, the recent COVID-19 pandemic highlighted the importance of health risks and raised concerns among investors on how companies respond to them (Dhar & Bose, 2022). The geopolitical events in Europe, the energy crisis, and the turmoil in the global financial markets, also create a new environment within which SRIs and ESG investments should be reconsidered. Therefore, more research to explore the effect of these new conditions is needed. Moreover, SRI/ESG information and data should become more comprehensive. Validation is also required to compare what is stated by companies and mutual funds regarding their SRI/ESG policies, and their actual activities. Finally, strengthening the regulation could improve transparency and set a common set of principles for all market participants.

References

Allevi, E., Basso, A., Bonenti, F., Oggioni, G., & Riccardi, R. (2019). Measuring the environmental performance of green SRI funds: A DEA approach. *Energy Economics, 79*, 32–44. https://doi.org/10.1016/j.eneco.2018.07.023

Ballestero, E., Bravo, M., Perez-Gladish, B., Arenas-Parra, M., & Pia-Santamaria, D. (2012). Socially responsible investment: A multicriteria approach to portfolio selection combining ethical and financial objectives. *European Journal of Operational Research, 216*(2), 487–494.

Basso, A., & Funari, S. (2014). Constant and variable returns to scale DEA models for socially responsible investment funds. *European Journal of Operational Research, 235*(3), 775–783. https://doi.org/10.1016/j.ejor.2013.11.024

Berry, T. C., & Junkus, J. C. (2013). Socially responsible investing: An investor perspective. *Journal of Business Ethics, 112*(4), 707–720. https://doi.org/10.1007/s10551-012-1567-0

Bilbao-Terol, A., Arenas-Parra, M., & Cañal-Fernández, V. (2012a). A fuzzy multi-objective approach for sustainable investments. *Expert Systems with Applications, 39*, 10904–10915.

Bilbao-Terol, A., Arenas-Parra, M., & Cañal-Fernández, V. (2012b). Selection of socially responsible portfolios using goal programming and fuzzy technology. *Information Sciences, 189*, 110–125.

Bollen, N. P. B. (2007). Mutual fund attributes and investor behavior. *Journal of Financial and Quantitative Analysis, 42*(3), 683–708. https://doi.org/10.1017/S0022109000004142

Candelon, B., Hasse, J.-B., & Lajaunie, Q. (2021). ESG-washing in the mutual funds industry? *From Information Asymmetry to Regulation. Risks, 9*(11), 199. https://doi.org/10.3390/risks9110199

Chen, X., & Scholtens, B. (2018). The urge to act: A comparison of active and passive socially responsible investment funds in the United States. *Corporate Social Responsibility and Environmental Management, 25*(6), 1154–1173. https://doi.org/10.1002/csr.1529

Climent, F., & Soriano, P. (2011). Green and good? The investment performance of US environmental mutual funds. *Journal of Business Ethics, 103*(2), 275–287. https://doi.org/10.1007/s10551-011-0865-2

Cortez, M. C., Silva, F., & Areal, N. (2012). Socially responsible investing in the global market: The performance of US and European funds. *International Journal of Finance & Economics, 17*(3), 254–271. https://doi.org/10.1002/ijfe.454

Dhar, S., & Bose, I. (2022). Victim crisis communication strategy on digital media: A study of the COVID-19 pandemic. *Decision Support Systems, 113830*. https://doi.org/10.1016/j.dss.2022.113830

Dorfleitner, G., Kreuzer, C., & Laschinger, R. (2021). How socially irresponsible are socially responsible mutual funds? *A Persistence Analysis. Finance Research Letters, 43*, 101990. https://doi.org/10.1016/j.frl.2021.101990

Friedman, M. (2007). The social responsibility of business is to increase its profits. In W. C. Zimmerli, M. Holzinger, & K. Richter (Eds.), *Corporate Ethics and Corporate Governance* (pp. 173–178). Springer Berlin Heidelberg. https://doi.org/10.1007/978-3-540-70818-6_14

Gangi, F., Varrone, N., Daniele, L. M., & Coscia, M. (2022). Mainstreaming socially responsible investment: Do environmental, social and governance ratings of investment funds converge? *Journal of Cleaner Production, 353*, 131684. https://doi.org/10.1016/j.jclepro.2022.131684

García-Melón, M., Pérez-Gladish, B., Gómez-Navarro, T., & Mendez-Rodriguez, P. (2016). Assessing mutual funds' corporate social responsibility: A multistakeholder-AHP based methodology. *Annals of Operations Research, 244*(2), 475–503. https://doi.org/10.1007/s10479-016-2132-5

Gasser, S. M., Rammerstorfer, M., & Weinmayer, K. (2017). Markowitz revisited: Social portfolio engineering. *European Journal of Operational Research, 258*(3), 1181–1190. https://doi.org/10.1016/j.ejor.2016.10.043

Geczy, C. C., Stambaugh, R. F., & Levin, D. (2021). Investing in socially responsible mutual funds. *The Review of Asset Pricing Studies, 11*(2), 309–351. https://doi.org/10.1093/rapstu/raab004

Hawn, O., Chatterji, A. K., & Mitchell, W. (2018). Do investors actually value sustainability? New evidence from investor reactions to the Dow Jones Sustainability Index (DJSI). *Strategic Management Journal, 39*(4), 949–976. https://doi.org/10.1002/smj.2752

Henke, H.-M. (2016). The effect of social screening on bond mutual fund performance. *Journal of Banking & Finance, 67*, 69–84. https://doi.org/10.1016/j.jbankfin.2016.01.010

Hong, H., & Kacperczyk, M. (2009). The price of sin: The effects of social norms on markets. *Journal of Financial Economics, 93*(1), 15–36. https://doi.org/10.1016/j.jfineco.2008.09.001

Junkus, J., & Berry, T. D. (2015). Socially responsible investing: A review of the critical issues. *Managerial Finance, 41*(11), 1176–1201. https://doi.org/10.1108/MF-12-2014-0307

Lapanan, N. (2018). The investment behavior of socially responsible individual investors. *The Quarterly Review of Economics and Finance, 70*, 214–226. https://doi.org/10.1016/j.qref.2018.05.014

Levering, R., & Moskowitz, M. (1998). The 100 best companies to work for in America. *Fortune, 137*, 84–95.

Matallín-Sáez, J. C., Soler-Domínguez, A., Navarro-Montoliu, S., & Mingo-López, D. V. (2022). Investor behavior and the demand for conventional and socially responsible mutual funds. *Corporate Social Responsibility and*

Environmental Management, 29(1), 46–59. https://doi.org/10.1002/csr.2172

Muñoz, F., Vargas, M., & Marco, I. (2014). Environmental mutual funds: Financial performance and managerial abilities. *Journal of Business Ethics, 124*(4), 551–569. https://doi.org/10.1007/s10551-013-1893-x

Nofsinger, J., & Varma, A. (2014). Socially responsible funds and market crises. *Journal of Banking & Finance, 48*, 180–193. https://doi.org/10.1016/j.jbankfin.2013.12.016

Olmedo, E. E., Torres, M. J. M., & Izquierdo, M. A. F. (2010). Socially responsible investing: Sustainability indices, ESG rating and information provider agencies. *International Journal of Sustainable Economy, 2*(4), 442. https://doi.org/10.1504/IJSE.2010.035490

Pastor, L., Stambaugh, R., & Taylor, L. (2019). *Sustainable investing in equilibrium* (No. 26549). National Bureau of Economic Research. https://doi.org/10.3386/w26549

Pérez-Gladish, B., Rodríguez, P. M., & M'zali. (2013). Mutual funds efficiency measurement under financial and social responsibility criteria. *Journal of Multi-Criteria Decision Analysis, 20*(3–4), 109–125. https://doi.org/10.1002/mcda.1494

Popescu, I.-S., Hitaj, C., & Benetto, E. (2021). Measuring the sustainability of investment funds: A critical review of methods and frameworks in sustainable finance. *Journal of Cleaner Production, 314*, 128016. https://doi.org/10.1016/j.jclepro.2021.128016

Raghunandan, A., & Rajgopal, S. (2022). Do ESG funds make stakeholder-friendly investments? *Review of Accounting Studies*. https://doi.org/10.1007/s11142-022-09693-1

Renneboog, L., Ter Horst, J., & Zhang, C. (2008). Socially responsible investments: Institutional aspects, performance, and investor behavior. *Journal of Banking & Finance, 32*(9), 1723–1742. https://doi.org/10.1016/j.jbankfin.2007.12.039

Riedl, A., & Smeets, P. (2017). Why do investors hold socially responsible mutual funds? *The Journal of Finance, 72*(6), 2505–2550. https://doi.org/10.1111/jofi.12547

Sandberg, J., Juravle, C., Hedesström, T. M., & Hamilton, I. (2009). The heterogeneity of socially responsible investment. *Journal of Business Ethics, 87*(4), 519–533. https://doi.org/10.1007/s10551-008-9956-0

Schueth, S. (2003). Socially responsible investing in the United States. *Journal of Business Ethics, 43*(3), 189–194. https://doi.org/10.1023/A:1022981828869

Shefrin, H., & Statman, M. (1985). The disposition to sell winners too early and ride losers too long: Theory and evidence. *The Journal of Finance, 40*(3), 777–790. https://doi.org/10.1111/j.1540-6261.1985.tb05002.x

Steuer, R. E., Qi, Y., & Hirschberger, M. (2007). Suitable-portfolio investors, nondominated frontier sensitivity, and the effect of multiple objectives on standard portfolio selection. *Annals of Operations Research, 152*, 297–317. https://doi.org/10.1007/s10479-006-0137-1

Townsend, B. (2020). From SRI to ESG: The origins of socially responsible and sustainable investing. *The Journal of Impact and ESG Investing, 1*(1), 10–25. https://doi.org/10.3905/jesg.2020.1.1.010

Utz, S., & Wimmer, M. (2014). Are they any good at all? A financial and ethical analysis of socially responsible mutual funds. *Journal of Asset Management, 15*(1), 72–82. https://doi.org/10.1057/jam.2014.8

Utz, S., Wimmer, M., Hirschberger, M., & Steuer, R. E. (2014). Tri-criterion inverse portfolio optimization with application to socially responsible mutual funds. *European Journal of Operational Research, 234*(2), 491–498. https://doi.org/10.1016/j.ejor.2013.07.024

Utz, S., Wimmer, M., & Steuer, R. E. (2015). Tri-criterion modeling for constructing more-sustainable mutual funds. *European Journal of Operational Research, 246*(1), 331–338. https://doi.org/10.1016/j.ejor.2015.04.035

van Duuren, E., Plantinga, A., & Scholtens, B. (2016). ESG integration and the investment management process: Fundamental investing reinvented. *Journal of Business Ethics, 138*(3), 525–533. https://doi.org/10.1007/s10551-015-2610-8

Vo, N. N. Y., He, X., Liu, S., & Xu, G. (2019). Deep learning for decision making and the optimization of socially responsible investments and portfolio. *Decision Support Systems, 124*, 113097. https://doi.org/10.1016/j.dss.2019.113097

Widyawati, L. (2020). A systematic literature review of socially responsible investment and environmental social governance metrics. *Business Strategy and the Environment, 29*(2), 619–637. https://doi.org/10.1002/bse.2393

Firm ESG Practices and the Terms of Bank Lending

Mingying Cheng and Iftekhar Hasan

1 Introduction

Over the past two decades, environmental, social, and governance (ESG) issues have attracted the attention of shareholders, institutional investors, financial analysts, debtholders and regulators (Amiraslani et al., 2021; Becchetti et al., 2012; Dyck et al., 2019; Flammer, 2021; Hasan et al., 2017; Krueger et al., 2020; Ioannou & Serafeim, 2015; SEC, 2022), and the dot-com bubble and the global financial crisis of 2008 were catalysts for even greater interest in socially responsible investing (SRI). Market participants blamed financial institutions for the financial crisis because

M. Cheng · I. Hasan (✉)
Fordham University, New York, NY, USA
e-mail: ihasan@fordham.edu

M. Cheng
e-mail: mcheng21@fordham.edu

I. Hasan
Bank of Finland, Helsinki, Finland

University of Sydney, Sydney, NSW, Australia

C. Gaganis et al. (eds.), *Sustainable Finance and ESG*,
Palgrave Macmillan Studies in Banking and Financial Institutions,
https://doi.org/10.1007/978-3-031-24283-0_5

of their lack of governance and narrow focus on short-term profits. At the same time, investors' increasing awareness of issues related to ESG highlight the risks of ignoring them. Thus, they demand ESG products and consciously incorporate social responsibility into their investment decisions.[1]

High-profile ESG-related controversies, such as Enron financial reporting, Volkswagen emissions, and Facebook data privacy scandals, have further brought ESG risks into the spotlight. As of the end of 2021, more than 3,800 institutional investors have signed on to the United Nations-supported Principles of Responsible Investment (PRI), a commitment to including environmental, social, and governance factors in investment decision and ownership. Assets under management for these investors increased from US$6.5 trillion in 2006 to over US$ 121 trillion in 2021, which represents a more than 1700 percent growth rate. Moreover, in his annual letter to CEOs, BlackRock's CEO Larry Fink highlighted the long-term economic benefits to companies of sustainability issues and, as the world's largest asset manager, reemphasized the stance of prioritizing sustainable investing.[2]

As pivotal intermediaries in financial markets, banks have begun to treat ESG as a priority. Responding to pressure from investors and regulators to address ESG-related risks, a growing number of banks have committed to support a transition to a more socially responsible economy. At the same time, banks are developing their own sustainability strategies, with European banks in the forefront and many US banks following closely behind. Mark Carney, the governor of the Bank of England, warned in 2015 that climate change has become a financial risk and a potential threat to overall financial-market stability. On the one hand, banks have pledged to assist in a socially responsible economic transformation. For example, as of April 2022, banks with more than $81 trillion, representing 45 percent of global banking assets, have signed up for the United Nations Principles for Responsible Banking and have agreed to align their lending and investment practices with the UN's Sustainable Development Goals and the Paris Climate Agreement.[3] For example, in 2021, the Bank of America, announced a goal of deploying $1 trillion through its

[1] https://www.responsible-investor.com/ri-landscape-news1/.

[2] The Power of Capitalism, Larry Fink's Letter to CEOs, 2022.

[3] https://www.unepfi.org/banking/bankingprinciples/prbsignatories/.

Environmental Business Initiative to reduce carbon-intensive lending and reach the net-zero goal by 2030, thus supporting a sustainable economy.[4] On the other hand, banks assess borrowers' ESG factors in financing contracts and create innovative loans to address potential ESG risks (e.g., sustainability-linked loans and green loans). Specifically, 67 percent of banks screen their loan portfolios for ESG risks and avoid those firms with high risks, leading to greater due diligence (Fitch, 2020). ESG loans such as sustainability-linked loans, where banks provide borrowers with a lower interest rate if they meet their sustainability targets or improve their metrics, have increased in popularity in recent years. According to Kim et al. (2022), more than 12 percent of global bank lending comprises ESG loans. All these factors suggest that banks have embedded ESG issues in their lending relationships to shift borrowers into more sustainable business models.

In this chapter, we review the recent literature on the interactions between borrowers' ESG practices and banking relationships. We first discuss the underlying economic theories regarding why banks change their lending terms contingent on borrowers' ESG profiles. Next, we examine the empirical literature that explores various channels through which firms' ESG activities affect banks' lending decisions, including loan interest rates, maturity, and collateral requirements. We also discuss how banks may affect borrowers' ESG policies and investments through lending relationships. We provide empirical evidence to show the value relevance of ESG risks for banks and investigate the learning process between banks and borrowers. We find that banks learn by observing the ESG behaviors of responsible borrowers and consequently improve their own ESG scores.

2 What Motivates Banks to Consider ESG as a Relevant Issue?

As the world's economies embracefinancial sustainability as a new business strategy, banks find themselves in a unique leadership position that can allocate the capital required and coordinate with multiple stakeholders to transition to a more sustainable and inclusive economy. Given their

[4] https://newsroom.bankofamerica.com/content/newsroom/press-releases/2021/04/bank-of-america-increases-environmental-business-initiative-targ.html.

influence and reach, banks must be involved in mitigating ESG-related risks and their potential impacts. In this section, we discuss several factors that drive banks to incorporate ESG into their practices. Specifically, we identify why banks consider ESG as a relevant issue for financial reasons, reputational risks, and emerging regulatory pressure.

ESG and Credit Risk

Why do banks care about borrowers' ESG performance? Research studies have documented that firms' ESG assessments provide valuable insights, both direct and indirect, into current and future financial performance and investment returns (Edmans, 2011; Kölbel et al., 2017; Lins et al., 2017). According to the Bank of America, ESG controversies wiped out more than $500bn market value from S&P 500 firms, and 90 percent of bankruptcies between 2005 and 2015 were firms with below-average environmental and social performance for the prior five years.[5] A well-established notion in the literature is that bank loans are an important financing source for companies, including for large public firms (Beck et al., 2008; Houston & James, 1996).Credit risk or default risk is the largest threat for banks. When pricing a loan, banks evaluate borrowers' credit risks and their profitability to repay it (Hasan et al., 2014; Sufi, 2009). Banks possess unique access to borrowers' ESG information and can effectively monitor their progress on those issues (Fama, 1985; James, 1987) and are motivated to incorporate borrowers' ESG performance because firms can "do well by doing good," which leads to lower credit risk. Specifically, companies focusing on outperforming ESG activities have enhanced corporate profits and long-term value by mitigating potential risks, including systematic, supply-chain, litigation, reputational, and regulatory risks.

ESG activities create competitive advantages that provide downside protection in volatile markets. For example, Bénabou and Tirole (2010) suggest that ESG can override managers' myopic decisions and protect firms from negative externalities from policymakers and market failures, such as lobbying, jurisdictional territoriality, market inefficiency from poor environmental information, and high transaction costs. In other

[5] Bank of America "10 reasons to care about environmental, social and governance (ESG)": https://about.bankofamerica.com/assets/pdf/BofA_ESG-10-reasons-you-should-care-about-ESG-Investing.pdf, January 7, 2020.

words, ESG creates long-term future benefits for firms. Similarly, Lins, Servaes, and Tamayo (2017) find that firms with strong ESG profiles have higher resilience during crisis periods, suggesting ESG activities can protect shareholder value against economic downturns. Relatedly, firms' ESG strategies differentiate them from competitors and can reduce price elasticity of demand, resulting in decreased systematic risks (Albuquerque et al., 2019). In other words, ESG-outperforming firms are less correlated with business cycles and can thus maintain stable cash flows and higher profit margins, thus leading to increased firm value.

Strong ESG profiles can also avoid potential costs from supply-chain disruption, litigation, and regulation as a result of trusted and loyal customers, satisfied employees, and clean and safe operations. First, customers take firms' ESG activities into account when making purchase decisions (Sen & Bhattacharya, 2001). Servaes and Tamayo (2013) document that a firm's good ESG policies can enhance customer awareness and loyalty, thereby increasing cash flows and firm value. In a similar spirit, Dai et al. (2021) focus on a unilateral effect on ESG engagement from customers to suppliers in global supply chains. They provide evidence that firms with better ESG performance generate firm value through improved operational efficiency and increased future sales. Second, a firm's own ESG policies may help to recruit, attract, and retain talent. Edmans (2011) finds a positive relationship between employee satisfaction and long-run stock returns because happy employees are motivated and more productive. Moreover, employee-friendly firms have better financial performance because they are less likely to have employee layoffs, encouraging higher sales per employee (Li et al., 2021). Third, socially responsible firms face less litigation risk. Hong and Kacperczyk (2009) suggest that the reason firms in the "sin" industries, i.e., those related to alcohol, tobacco, and gaming, have higher expected returns than otherwise comparable firms is their greater scrutiny and litigation risk from norm-constrained investors. As they point out, tobacco companies bore significant litigation risk until 1997 when they reached a settlement with state governments. El Ghoul et al. (2011) find supporting evidence that firms can alleviate litigation risks by engaging with ESG practices to attract socially responsible investors, thereby reducing equity financing cost. Finally, reputational damage is a critical cost for firms' socially irresponsible activities. Firms with ESG misconduct citations are more exposed to stakeholder sanctions that lead to damaged reputation and increased financial risk (Becchetti & Manfredonia, 2022; Kölbel et al., 2017). Interestingly, based on a large sample of firms that violate environmental regulations, Karpoff et al. (2005) conclude that legal and regulatory penalties account for the major losses in firm share value, rather than reputational costs.

ESG and Bank Value Relevance

While banks evaluate borrowers' ESG profiles to prevent potential credit risks, their own ESG activities are also relevant to their reputation and market value. Banks have an immense and yet largely untapped opportunity to build and develop new markets that are more efficient and sustainable. Socially responsible banks can not only increase overall social benefits but can also enjoy greater financial performance. For example, ethical banks can promote social welfare by investing in ethical projects, resulting from reduced financial friction (Barigozzi & Tedeschi, 2015). Chih et al. (2010) demonstrated that socially responsible banks obtain a competitive advantage in open markets. Moreover, banks with better ESG activities have positive financial performance in terms of returns on assets and on equity, as well as net interest and non-interest income (Wu & Shen, 2013).

Given increased attention to ESG issues, institutional investors, financial analysts, debtholders, and regulators have incorporated firms' ESG performance into their decision-making when assessing market value (Amiraslani et al., 2021; Becchetti et al., 2012; Flammer, 2021; Ioannou & Serafeim, 2015; Krueger et al., 2020; SEC, 2022). If banks do not prioritize ESG issues, they will face increased scrutiny from government and society. Specifically, banks are open to potential adverse ESG impacts that arise from financial services to borrowers in sensitive industries. The idea of boycotting and disinvesting in "dirty" industries, such as oil, gas, and coal operations has gained momentum and financing those industries creates a stigma for banks. For example, in the USA, the Biden administration has recently restricted bank financing of new carbon-intensive fossil-fuel projects overseas. A bank's own non-ESG-compliant behaviors can induce reputational risks and triggering financial damage. Below, we provide new evidence on ESG reputational risks for banks, showing that ESG misconduct is value-relevant to their stock price, suggesting ESG-related risks are becoming increasingly important factors for financial institutions.

3 What and How ESG Factors Affect Bank Lending Terms

Environmental factors and bank loans

A nascent literature in finance provides theoretical and empirical evidence that banks factor borrowers' ESG profiles into loan pricing. Notably, many studies focus on the challenges stemming from climate-change risks.[6] Climate change has a potentially devastating long-term effect on future economic activities (Stern, 2007). Such risks are becoming increasingly relevant, with a high probability that they will materialize in the near future (Ilhan et al., 2021; Krueger et al., 2020). Broadly speaking, climate risks can be categorized into two types: physical risk and transition risk (Giglio et al., 2021). Physical risks are associated with costs and damage resulting directly from extreme weather events or natural disasters, for example, the threat of disruptions to food producers from droughts (Hong et al., 2019). Transition risks account for potential disruptions, such as from government regulation and shifting consumer preferences for environmentally friendly products, to firms' operations and business models when transforming into a greener and lower-carbon economy. For example, to combat global warming, in 2016, California Senate Bill 32 mandated the reduction of greenhouse gas emissions to 40 percent below 1990 levels by 2030.[7] Different industries may be disproportionately exposed to climate risks. For instance, the coastal real estate industry faces higher physical risks from rising sea levels, while the coal industry is more likely to suffer from transition risks such as carbon taxes.

Several studies have shown that banks view climate change as a relevant risk factor and incorporate it into different dimensions of their loan contracts. On the one hand, studies have examined the impact of physical risks on credit lines in bank loans. Brown et al. (2021) show that unexpectedly extreme weather events significantly reduce firm-level cash flows, leading to lower liquidity. To prevent default risk, banks charge higher interest rates, extend loans with shorter maturity, increase the probability of secured loans, and have variable interest rates. Similarly, Javadi and Masum (2021) and Correa et al. (2022) find consistent adverse effects of

[6] Giglio et al. (2021) discuss a comprehensive literature review on the impact of climate change across different assets.

[7] https://leginfo.legislature.ca.gov/faces/billTextClient.xhtml?bill_id=201520160SB32.

climate-change risks on the cost of bank loans. Specifically, firms located in areas more exposed to hazards such as drought and hurricanes pay higher spreads on bank loans. In terms of home loans, banks charge higher interest rates for mortgages on properties with higher risk of sea-level rise (Nguyen, Ongena, Qi, & Sila, 2021) and are more likely to initiate securitized mortgages in areas threatened by flooding (Ouazad & Kahn, 2021). On the other hand, the literature has shown that banks incorporate transition risks into commercial lending. Beyene, et al. (2020) find that firms with high levels of fossil-fuel reserves are penalized with higher yield spreads with growing attention to pro-environmental regulations and public policy. Ivanov, Kruttli, Sumudu and Watugala (2021) focus on the transition risk stemming from California SB 32 on greenhouse gas (GHG) emissions and document that affected high-emission firms have shorter loan maturities, limited access to permanent forms of bank financing, increased interest rates, and a higher participation of shadow banks in their lending syndicates. Banks became increasingly concerned about the default risk for firms with higher environmental liability after the 2008 Apex Oil settlement, and consequently require higher loan spread for such firms (Chen et al., 2022). Kacperczyk and Peydró (2021) show that environmentally committed banks allocate more credits to low-emission (greener) firms as they recognize those firms as having lower financial risks. Finally, investors demand a higher cost of capital for firms with poor environmental performance due to high regulatory concerns (Chava, 2014). Interestingly, current evidence highlights significant changes in bank lending after the Paris Climate Agreement of 2015. Specifically, banks began to price borrowers' environmental performance following adoption of the agreement (Degryse et al., 2022; Beyene et al., 2020; Ehlers et al., 2022) and bank lending has been reduced (Reghezza et al., 2021).

Social Factors and Bank Loans

The COVID-19 pandemic and the events of the #MeToo and "Black Lives Matter" movements have intensified discussion about the importance of incorporating the "S" (social) factor of ESG in corporate management. Firms face mounting pressure from investors, consumers, and regulators to consider social issues related to workplace safety, racial inclusion, and gender equality and numerous firms have pledged to address such concerns. For example, Goldman Sachs launched a

$10-million Fund for Racial Equity to support the vital work of encouraging organizations to focus on racial injustice, structural inequity, and economic disparity.[8] In contrast to the climate/environmental risks that can evolve over the long term, social risks such as the #MeToo movement can backfire instantaneously and are even amplified by social media. Banks can be indirectly exposed to financial losses or reputational damage if their borrowers are affected by such socially negative consequences. Therefore, banks must identify and integrate borrowers' social practices in their contracting terms.

Francis et al. (2019) find strong evidence that employee-friendly firms are better financially positioned to repay their debt and, thus, obtain lower cost of bank loans in both price and non-price terms. Qian et al. (2021) show similar results and theorize that banks treat firms with better employment policies as more trustworthy, because better treatment of employees implies both stronger operating ability and good intent toward creditors. Hasan et al. (2017) find that firms headquartered with a high level of social capital obtain cheaper loans because banks perceive social capital as an external pressure to harness opportunistic firm behaviors in debt contracting. Many existing studies have focused on "E" (environmental) and "G" (governance) issues, which has overshadowed the "S" perspective. As social issues evolve and gain attention, more studies are needed to investigate the potential social risks of bank lending.

Governance Factors and Bank Loans

Another stream of the literature focuses on the role of corporate governance in bank loan contracting, especially after a series of accounting-reporting scandals such as those involving Enron and Arthur Andersen. These studies are motivated by the effects of asymmetric information and agency risks on the cost of capital (Rajan & Winton, 1995). Lenders are more likely to charge higher interest rates and tighten non-price debt terms when firms with a high level of information-asymmetry and moral-hazard concerns, because lenders must compensate for greater credit risk and the need for greater monitoring efforts (Bhojraj & Sengupta, 2003). On the one hand, weak internal control is prone to both intentional reporting biases and unintentional accounting errors, which leads

[8] https://www.goldmansachs.com/citizenship/fund-for-racial-equity/

to higher cost of bank loans (Kim et al., 2011). On the other hand, banks grant more favorable loan terms to firms with strong takeover defense, since firms with strong shareholder rights have lower takeover vulnerability and subsequent financial risk (Chava and Roberts, 2008). Prior studies have also shown that an effective board of directors can significantly reduce information asymmetry, agency problems, and default risk (Core et al., 1999; Klein, 2002). Several papers have shown the benefits of effective boards on bank loans and find supporting evidence that firms with high-quality boards are recognized by banks and rewarded with favorable terms when pricing bank-loan contracts. An effective board can address agency problems and improve information asymmetry between the firm and outsiders (Fama & Jensen, 1983). Francis et al. (2012) document that that a more independent board is associated with lower interest rates and fewer restrictive collateral, covenants, and performance-pricing provisions. Similarly, Ge et al. (2012) find consistent evidence in an international context where banks provide cheaper loan contracts to firms with better internal governance, especially for those firms incorporated in countries with strong legal institutions. Moreover, board diversity reduces the cost of bank loans, especially female board representation (Karavitis et al., 2021), because female board members typically demand greater reporting transparency and monitoring of managers' actions (Adams & Ferreira, 2009). Accordingly, gender diversity can improve the quality of boards by increasing disclosure of more firm-specific information (Gul et al., 2011), leading to more in-depth knowledge for banks to assess the credit of potential borrowers. Furthermore, banks grant favorable credit terms to female-CFO-led firms because female CFOs are perceived as risk-averse and can alleviate credit risks (Francis et al., 2013).

Finally, banks reward transparent firms for ESG activities, which means that firms with high levels of ESG disclosure can obtain a lower cost of debt, for both EU firms (Eliwa, Aboud, and Saleh, 2021) and US firms (Degryse, Goncharenko, Theunisz, and Vadasz, 2022). Goss and Roberts (2011) find that banks charge significantly higher loan spreads to firms with below-average corporate social responsibility (CSR).

4 What are the Real Consequences of ESG Lending?

While extensive of evidence has shown that banks incorporate borrowers' ESG performance into loan decisions, what are the sustainable and financial outcomes of such bank lending? In other words, can banks and borrowers really "walk the ESG talk"? It is highly plausible that both borrowers and lenders are motivated to engage in ESG lending for greenwashing purposes, whereby they use ESG lending to gain legitimacy from stakeholders, but do not follow up on their commitments (Marquis et al., 2016; Raghunandan & Rajgopal, 2021). For example, on the one hand, borrowers strategically improve their ESG image in a short time period to attract cheaper loans from banks. Shin (2021) finds that borrowers endeavor to improve their ESG performance while seeking a loan; however, after receiving the loan, they reduce their ESG efforts, which indicates greenwashing around loan issuance. Similarly, Kim et al. (2022) document a decrease in ESG performance after loan origination, and Kacperczyk and Peydró (2021) show that bank loans have no effect on subsequent reduction in carbon emissions for environmentally concerned firms. On the other hand, banks also engage in social greenwashing. Socially responsible banks are expected to generate social welfare by providing credit to local communities, especially to underprivileged neighborhoods, with the passage of Community Reinvestment Act (CRA) of 1977. Surprisingly, however, Basu et al. (2022) have shownthat high-ESG-perceived banks extend fewer home-purchase loans in poor neighborhoods than do low ESG banks. Moreover, from the corporate-governance perspective, previous studies have viewed banks as "monitors" and "insiders" for borrowers (Fama, 1985; James, 1987). Stronger bank-firm lending relationships induce better monitoring and improve borrowers' corporate governance (Dass & Massa, 2011). The monitoring role of banks also enhances borrowing firms' disclosure quality by decreasing the degree of earnings management (Ahn & Choi, 2009). Houston and Shan (2022) propose a novel disciplinary mechanism in the banking relationship. Specifically, they suggest that, since it is costly for borrowers to switch lenders, banks can exert a monitoring effect on borrowers' ESG practices and enhance their ESG performance through "fear of subsequent exit."

Furthermore, some studies have shed light on the financial consequences for ESG-incompetent borrowers. Kacperczky and Peydró (2021) investigate the firm-level real effects from an environmental aspect, and point out that polluting firms experienced deleverage, indicated by lower leverage and reduced asset size. Nevertheless, no significant changes in environmental expenditures are apparent for those brown firms. Consistently, bank-dependent firms that are exposed to extreme hazard events tend to decrease their capital expenditure and increase cash holdings as concerns around climate-change risks grow (Correa et al., 2022).

5 New Empirical Evidence on the Impact of ESG from the Bank Perspective

We provide some empirical evidence to gain further insight into bank relationships with ESG. We first show the value-relevance of banks' ESG exposure and find, specifically, that bank ESG risks have a moderating effect on earnings. Moreover, to complement the findings of Houston and Shan (2022) that banking relationships act as a transmission mechanism whereby borrowers are influenced by socially responsible lending and promote corporate ESG behavior, we show that banks learn from the ESG behaviors of responsible borrowers and consequently further improve their own ESG performance.

We collect data from multiple sources over the period from 2000 to 2020. We obtain ESG scores from MSCI, which evaluates and is considered as the largest data provider to the investment community (Avramov et al., 2022; Ferrell, Hao and Renneboog, 2016); ESG risk data comes from RepRisk; syndicated loan data from Thomson Reuters' LPC DealScan; and lenders' and borrowers' financial data from Compustat. To test the value relevance of ESG risk on banks, we merge *RepRisk* with Compustat. Our sample period is limited by the RepRisk database that started in 2007. For the analysis of bank relationships and ESG performance, we merge Thomson Reuters' LPC DealScan and MSCI databases. It is possible that borrowers' country characteristics affect lenders' incentives to monitor, which can lead to different loan pricing. For example, borrowers from countries with a high level of property-rights protection and democracy enjoy cheaper loans (Bae & Goyal, 2009; Delis et al., 2019) and vice versa. To avoid potential country-level confounding factors, we limit our sample to US banks and US borrowers.

The Value Relevance of ESG Risks for Banks

Extensive research has documented the importance of accounting fundamentals to firm valuation (Barth et al. 2001; Beaver 1968; Collins et al., 1997). However, more recent studies have revealed a decline in the value relevance of accounting information (Barth et al., 2022; Balachandran & Mohanram, 2011), leading to the exploration of the value relevance of non-financial information, especially ESG information (Lourenço et al., 2014). As discussed earlier, ESG risks are important to capital market participants and have a critical impact on firms' market value, stakeholder impressions, and reputation. The question of whether ESG risks are incremental to the value relevance of earnings and book value of equity is worthy of investigation. Focusing on ESG risks, rather than on activities or disclosures, is likely to capture more relevant and more interpretable information for investors since these ESG issues are more likely to directly impact bottom-line profitability, as opposed to, for example, positive ESG performance. Therefore, we expect ESG risks will provide incremental information content for the relationship between accounting fundamentals and the expected value of future cash flows.

To test our hypothesis, we utilize a news-event data source, *RepRisk*, that captures bank exposure to ESG-related risks and test whether these risks provide incremental information to the relationship between key accounting fundamentals and bank market values. RepRisk overcomes several concerns about ESG scores from the rating agencies regarding measurement issues (Berg, Koelbel, & Rigobon, 2019; Dimson et al., 2020), selective disclosure (i.e., greenwashing) (Marquis et al., 2016), and non-uniformity reporting (Eccles & Serafeim, 2013). RepRisk is a popular data source for media coverage of ESG incidents (Kölbel et al., 2017). It tracks firm-level daily ESG negative news since 2007 and evaluates each news event in two aspects: severity (the harshness of the incident) and reach (the influence or the readership of the source). Finally, based on the number of negative ESG news events and the severity and reach of the incidents, RepRisk assigns a current RepRisk Index (the *Current RRI*) to each firm at the monthly level. The *Current RRI* ranges from 0 (lowest) to 100 (highest). As an alternative measure for bank ESG risks, we use *RepRisk Rating* (*RRR*), which is a letter rating (AAA to D) calculated based on two factors: firm-level peak ESG risk and country-sector ESG risk. We convert the letter grade of *RRR* to a numerical scale to use it in regressions. Specifically, we convert the *RRR* to a scale of 1 to 10, where 1 represents the lowest risk "AAA" and 10 represents the highest risk "D." We collect financial fundamental information for banks identified with two-digit SIC codes 60–62 from Compustat. Our sample consists of 1,436 unique banks with 11,178 bank-year observations from 2007 to 2020.

As shown in Fig. 1, banks on average have much higher number of negative ESG incidents relative to borrowers, corresponding to the growing attention to ESG issues. To show the moderating effect of ESG risk exposure on a bank's financial fundamentals, we proceed in several ways. First, we plot the evolution of the bank's ESG risk exposure (Fig. 2) and the combined value relevance of its accounting fundamentals (Fig. 3). Figure 2 illustrates the trend in the bank's average ESG risk exposure over time, defined by the *RepRisk* percentages of ESG risks. While environmental exposure is quite low and stays flat, banks face increasing exposure to social and governance risks over the sample period. As expected, banks' own activities are less likely to have environmental risks, but they are scrutinized by the capital markets for lending money to environmentally irresponsible firms. After the 2007–2008 financial crisis, investors became aware of the importance of governance; thus, bank exposure to governance risks has increased dramatically from 0 to 11% since then. Bank exposure to social risks also presents an upward trend, corresponding to the growing attention to discrimination, equality, and justice issues. We then interact the bank's ESG risks with the standard value-relevance model, following Balachandran and Mohanram (2011). Specifically, we save the adjusted R-squared from yearly regressions of the Ohlson (1995) model, before and after the inclusion of ESG risks. The value of adjusted R-squared captures the extent to which financial fundamentals and incremental ESG risk information explain investors' beliefs about bank values. As shown in Fig. 3, the blank circle represents the adjusted R-squared values for the Ohlson (1995) model and the solid diamond represents the adjusted R-squared values for the inclusion of the moderating effect of ESG risks in the Ohlson (1995) model. Figure 3 illustrates that every solid diamond is above the blank circle, suggesting ESG risks provide incremental explanatory power to investors' valuation of banks. Specifically, the incremental explanatory power has increased since 2018, as the gap between the two groups becomes larger. Taken together, both Figs. 2 and 3 show that ESG risks are of growing importance for banks.

Second, we employ OLS regressions using the modified Ohlson (1995) value-relevance model following (Barth, Li, and McClure, 2022) and construct a specific sample with bank financial performance and ESG activities. Following Barth et al. (2022), our empirical specification is as below:

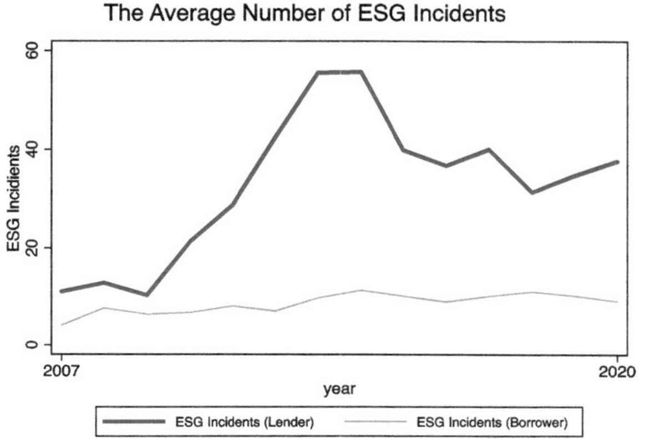

Fig. 1 Average number of ESG incidents covered by media

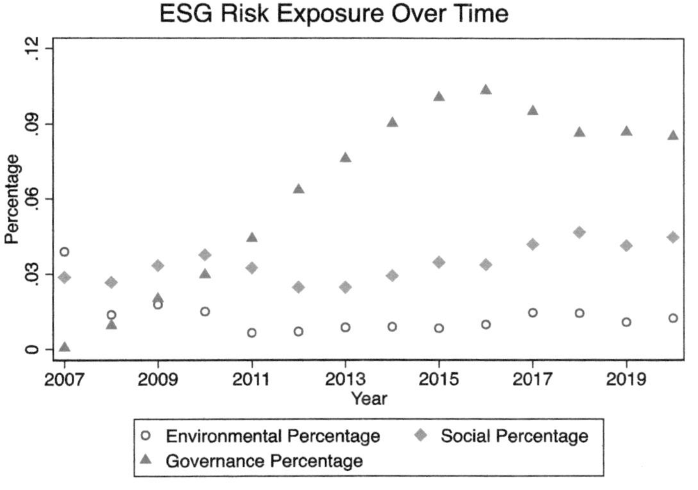

Fig. 2 Banks' mean environmental, social, and governance risk exposures over time

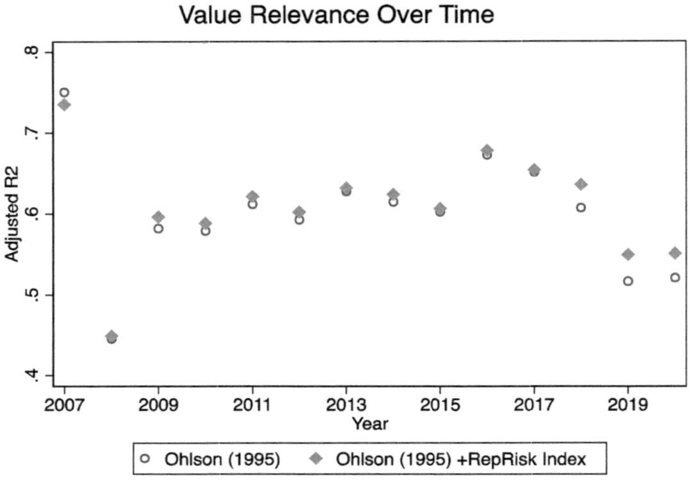

Fig. 3 Value relevance of earnings and the book values of equity over time

$$\begin{aligned}\text{StockPrice}_{it} = &\alpha_0 + \beta_1\text{EPS}_{it} + \beta_2\text{BVPS}_{it} + \beta_3\text{RepRiskIndex}_{it} \\ &+ \beta_4\text{EPS} \times \text{RepRiskIndex}_{it} + \beta_5\text{BVPS}_{it} \\ &\times \text{RepRiskIndex}_{it} + \beta_6\text{Controls}_{it} \\ &+ \text{Year}_{it} + \varepsilon_{it}\ (1)\end{aligned}$$

$StockPrice_{it}$ is the stock price of bank i 3 months following the end of fiscal year t. $RepRiskIndex_{it}$ is the annual average of ESG risk for bank i in year t. For controls, we include a vector of variables following Barth and Clinch (2009). Specifically, *Assets* is the total assets (AT); *OCF* is the operating cash flow (OANCF); *Cash* is the cash and short investment (CHE); *Dividends* is the dividends on common shares (DVC); *R&D* is the R&D expense (XRD); *Intangibles* is the intangible assets (INTAN); *Advertising* is the advertising expense (XAD); *Special Items* is the SPI; *OCI* is calculated as other comprehensive income (OCI) minus the prior year retained earnings (RE), plus the dividends on common shares (DVC) and minus the income before extraordinary items (IB); *Revenues* is the total revenue (REVT). All control variables are scaled by the total number of shares outstanding at the end of fiscal year t.

Table 1 Summary Statistics Panel A. The Value-RelevanceBanks' ESG value-relevance of Bank ESG RisksNMeanStd. DevMinMedianMaxStock

Price11,17824. Panel A reports the summary statistics. On average, firms' stock prices are $24.878, earnings per share are $1.443, and book value of equity are $18.148 per share. The RepRisk Index (RRI) has a mean of 3.450 and a standard deviation of 9.437, suggesting wide-spread heterogeneity in ESG risk exposure. The RepRisk Rating (RRR) ranges from 0 to 10, with an average value of 0.923 and a standard deviation of 1.673.

Table 2 presents the results. As shown in Table 2 column (1), we find evidence consistent with prior literature, where the stand-alone earnings

Table 1 Summary statistics

Panel A: The value-relevance of bank ESG risks

	N	*Mean*	*Std. Dev*	*Min*	*Median*	*Max*
Stock Price	11,178	24.878	34.579	0.005	15.450	419.250
RepRisk Index (RRI)	11,178	3.450	9.437	0.000	0.000	68.083
RepRisk Rating (RRR)	11,178	0.923	1.673	0.000	0.000	9.000
EPS	11,178	1.443	2.791	−15.647	1.054	17.388
BVPS	11,178	18.148	17.144	−10.167	13.871	121.437
Assets	11,178	170.919	165.529	0.000	133.482	994.931
OCF	11,178	2.728	4.628	−7.774	1.741	36.294
Cash	11,178	12.667	17.730	0.000	6.230	128.911
Dividends	11,178	0.562	0.774	0.000	0.314	5.177
R&D	11,178	0.008	0.089	0.000	0.000	3.325
Intangibles	11,178	3.411	7.769	0.000	0.756	82.392
Advertising	11,178	0.112	0.268	0.000	0.038	3.498
Special Items	11,178	−0.143	0.721	−10.155	0.000	1.803
OCI	11,178	−0.133	1.288	−10.857	−0.014	7.440
Revenues	11,178	12.664	21.210	0.000	7.815	262.149

Panel B: Banks ESG and lending relationships

Variable	*Obs*	*Mean*	*Std. Dev*	*Min*	*Median*	*Max*
ESG_Chg_Bank	3,303	0.012	1.752	−4.833	0.000	4.583
ESG_Diff	3,303	0.246	2.062	−6.000	0.167	6.000
Borrower_Chg	3,303	0.026	1.360	−5.000	0.000	5.000
ESG_Bank	3,303	2.185	1.448	0.000	2.250	6.000
lnPackageAmt	3,303	6.572	1.249	2.303	6.621	9.082
Firm Size	3,303	9.039	1.505	1.216	8.979	12.174
ROA	3,303	0.047	0.088	−1.891	0.047	0.183
Leverage	3,303	0.312	0.173	0.000	0.298	0.761
MTB	3,303	1.294	1.146	0.090	0.969	11.281

and book values of equity are associated with firm value (Barth et al., 1998; Beaver, 1968; Ohlson, 1995). In column (2), we examine the moderating effect of bank ESG risk on the impact of earnings and book value of equity on stock price. The coefficient of the interaction term, *EPS* × *RepRisk Index*, is 0.068 at the 5% significance level, suggesting that current bank ESG risk provides some incremental, forward-looking information content regarding market valuation. However, the coefficient of the interaction term, *BVPS* * *RepRisk Index*, is −0.009, which is not statistically significant. This finding means that current ESG risks for banks are not relevant for the ability of book value of equity to predict future cash flows. Accordingly, we rerun the model with the alternative measure for bank ESG risks, *RepRisk Rating* (*RRR*). As shown in column (3), our results hold. The coefficient of the *EPS* × *RRR*, is 0.452, at the 10% significance level, while the coefficient of the interaction term, *BVPS* × *RepRisk Index*, is −0.065 and is not statistically significant. Figure 4 plots the interaction effect between ESG risks and EPS on stock price. Basically, bank ESG risks can adjust the ability of current earnings to predict future cash flows but cannot modify the relevance of book values of equity to stock prices.

Banks' ESG and lending relationships

As Houston and Shan (2022) have documented, banking relationship is a novel disciplinary mechanism whereby banks prefer borrowers with ESG profiles that are similar to theirs. High ESG banks are sensitive to their reputations and thus want to avoid damage and public scrutiny due to lending relationships with socially irresponsible borrowers. Moreover, they provide evidence that lenders significantly influence the evolution of their borrowers' ESG profiles. Specifically, a one-standard-deviation increase in the distance between bank and borrower ESG ratings is associated with a 0.66 increase in the borrower's *RepRisk* rating over a two-year window centered on the loan package initiation date. As Houston and Shan (2022) have argued, banks have strong incentives to monitor and discipline borrowers from ESG misconduct, thus *RepRisk* perfectly suits their setting. In this chapter, we revisit this banking relationship and build on Houston and Shan (2022)'s study by investigating whether the ESG learning process occurs in reverse. To be specific, banks can closely observe their borrowers' ESG activities, such as ESG initiatives, management, and reporting practices and, in particular, banks with poor ESG

Table 2 OLS Regressions of *Stock Price* on *EPS*, *BVPS*, and *ESG Risk Exposure* and Controls

Dependent variable	*(1)* *Stock Price*	*(2)* *Stock Price*	*(3)* *Stock Price*
EPS	1.461***	1.167***	0.949***
	(0.330)	(0.315)	(0.298)
BVPS	0.970***	1.027***	1.068***
	(0.191)	(0.183)	(0.179)
RepRisk Index		0.031	
		(0.144)	
EPS * RepRisk Index		**0.068****	
		(0.029)	
*BVPS * RepRisk Index*		−0.009	
		(0.006)	
RRR			0.888
			(1.094)
EPS * RRR			**0.452****
			(0.177)
*BVPS * RRR*			−0.065
			(0.046)
Assets	−0.012	−0.012	−0.011
	(0.013)	(0.013)	(0.013)
OCF	0.127	0.124	0.137
	(0.097)	(0.097)	(0.097)
Cash	0.071	0.072	0.070
	(0.046)	(0.047)	(0.047)
Dividends	4.131**	4.010**	4.038**
	(1.834)	(1.818)	(1.825)
R&D	37.785*	37.740*	37.573*
	(19.310)	(19.310)	(19.206)
Intangibles	−0.051	−0.033	−0.026
	(0.201)	(0.195)	(0.189)
Advertising	−6.720	−6.598	−6.627
	(6.773)	(6.588)	(6.608)
Special Items	−1.161**	−1.163**	−1.221**
	(0.493)	(0.480)	(0.485)
OCI	0.208	0.239	0.172
	(0.257)	(0.254)	(0.253)
Revenues	0.415***	0.399***	0.391***
	(0.089)	(0.086)	(0.083)

(continued)

Table 2 (continued)

Dependent variable	*(1)* *Stock Price*	*(2)* *Stock Price*	*(3)* *Stock Price*
Constant	−6.077*	−6.127*	−6.653*
	(3.594)	(3.472)	(3.572)
Observations	11,178	11,178	11,178
Adjusted R-squared	0.865	0.867	0.867
FIRM FE	Yes	Yes	Yes
YEAR FE	Yes	Yes	Yes
Error cluster	Firm	Firm	Firm

This table reports the results from OLS regressions of stock prices on ESG risk disclosure, accounting fundamentals, and their interactions. The dependent variable is the stock price three months after the end of fiscal year *t*. BVPS is the book value per share calculated as book value of equity (CEQ) divided by total number of shares outstanding (CSHO) at the end of fiscal year *t*. *EPS* is earnings per share calculated as income before extraordinary items (IB) divided by the total number of shares outstanding at the end of fiscal year *t*. Each regression controls for firm and year fixed effects. Robust standard errors are clustered at firm level. *, **, *** Indicates statistical significance at 10%, 5%, and 1%, respectively

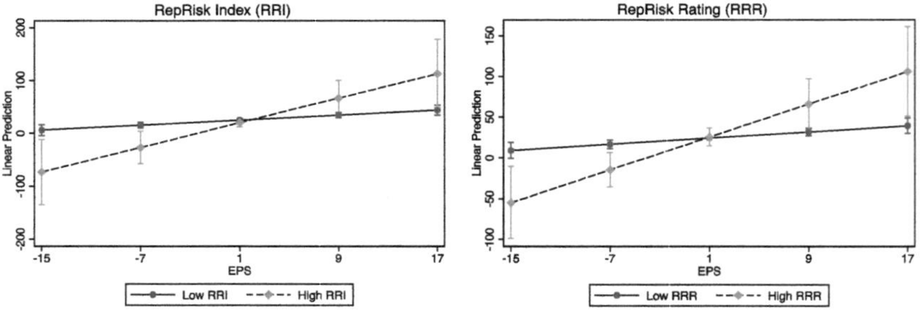

Fig. 4 Moderating effect of ESG risks on EPS

performance may be more incentivized to improve their ESG profiles and learn from borrowers with strong ESG ratings, therefore subsequently boosting their own ESG scores.

To explore this hypothesis, we employ a leading ESG ratings source provided by MSCI IVA database that has been extensively used in recent studies (Avramov et al., 2022; Ferrell, Hao and Renneboog, 2016). The MSCI research team assesses firms' ESG activities based on information collected from annual and corporate sustainability reports, nongovernmental organizations, and news events. Firms are rated on a scale ranging from AAA (best) to CCC (worst), relative to the performance of their industry peer groups. Following Ferrell, Hao and Renneboog (2016), we convert the MSCI's letter scale into a numeric scale (e.g., AAA is 6; AA is 5; ... and CCC is 0).

We acquire bank loan data from LPS DealScan that contains detailed information on borrowers and lenders, and loan characteristics such as loan spreads, size, maturity, type, purpose, collateral, and other loan terms. As Chava and Roberts (2008) estimated, about 60 percent of loan data from LPS DealScan is derived from SEC filings. We link borrowing firms to financial fundamentals from Compustat using the mapping provided by Michael R. Roberts (Chava & Roberts, 2008). Following prior studies, we focus on the lead arrangers in each loan as the lead banks establish and maintain an arm's-length relationship with the borrowers (Sufi, 2007). Specifically, we rely on the "LeadArranger-Credit" indication and, if the field shows "Yes," we identify the lender as a lead arranger (Bharath et al., 2009; Houston & Shan, 2022). For those facilities with multiple lead arrangers, we calculate the average of ESG ratings of lead-lenders in the syndicate. We further exclude from our sample borrowers in regulated and financial industries identified with two-digit SIC codes 40–45 and 60–64, following Ivashina (2009). Our final sample consists of 1,323 unique borrowing firms and 133 unique lending banks, resulting in 3,303 loan observations between 2000 and 2020.

To examine how banks observe borrowers' ESG activities and adjust their own ESG profiles, we estimate a regression model following Houston and Shan (2022):

$$\begin{aligned} \text{ESG\_ Chg\_ Bank}_{i,t-1,t+1} = & \beta \text{ESG\_ Diff}_{i,j,t-1} \\ & + \gamma \text{Borrower\_ Chg}_{j,t-1,t+1} \\ & + \lambda \text{ESG\_ Bank}_{i,t-1} + \delta \text{Controls}_{j,t} \\ & + \sum \text{Industry} + \sum \text{Year} + \varepsilon_{ijt} \quad (2) \end{aligned}$$

The unit of observation is the loan package. $ESG\_Chg\_Bank_{i,t-1,t+1}$ is our main outcome variable, calculated as the change in bank *i*'s ESG ratings over a two-year window, from one year before *t–1* to one year after the package initiation date *t + 1*. Our key independent variable is $ESG\_Diff_{i,j,t-1}$ that captures the difference between the borrower *j*'s and bank *i*'s ESG rating, measured one year before the package initiation date. $Borrower\_Chg_{j,t-1,t+1}$ controls for the change in a borrower's ESG score over the same two-year window. $ESG\_Bank_{j,t-1}$ accounts for the possibility of path-dependency concerns, where we compare banks with similar ex-ante (i.e., $t-1$) ESG profiles. For other control variables, we include loan and borrower characteristics that could potentially affect the bank's ESG ratings. At the loan-level, we control for the loan amount (*lnPackageAmt*) measured as the log of package amount in $millions. At the borrower-level, we include company size, ROA, leverage ratio, and market-to-book (*MTB*) ratio.

To account for unobservable variations from the loan supply side, we include Fama–French 12 industry and year fixed effects. ε_{ijlt} is the error term. We use two-way cluster standard errors at the lender and year level. All continuous variables are winsorized at 1% and 99% levels to remove outliers. Appendix 1 documents detailed definitions for each variable.

Table 1 Panel B reports the summary statistics. In our sample, we observe that *ESG_Chg_Bank* ranges from −4.833 to 4.583 with a mean value of 0.012. The difference between borrower and bank (*ESG_Diff*) varies from −6.000 to 6.000 with an average value of 0.246, and the change in borrowers' ESG rating (*Borrower_Chg*) has a minimum value of −5.000, a maximum value of 5.000, and a mean value of 0.026. On average, banks' ex-ante ESG score (*ESG_Bank*) is 2.185 with the standard deviation of 1.448. A loan package in the sample has an average size of 713.798 million and is 6.572 in logarithm. As for borrowers, the mean firm size is 8,424.347 million, and 9.039 in logarithm. On average, the ROA is 0.047, the leverage ratio is 0.312, and the MTB ratio is 1.294.

Table 3 column (1) reports the results based on the full sample. We find that the coefficient of *ESG_Diff* has a positive and significant effect on bank ESG change at the 10% significance level, suggesting that the larger gap between borrower and bank ESG rating is significantly related to the change in a bank's ESG rating over time. In terms of economic magnitude, a one-standard-deviation increase in *ESG_Diff* is associated with a 0.062 ($= 2.062 \times 0.030$) increase in the bank's ESG rating (*ESG_Chg_Bank*) over the two-year window. Equivalently, bank ESG rating is increased by 3.54% ($= 0.062/1.752$) of the standard deviation. Our finding indicates that, in general, banks improve their ESG ratings via the lending relationship.

Next, we explore the asymmetric influence of borrowers on banks. As Houston and Shan (2022) have documented, banks with relatively

Table 3 Evolution in banks' ESG ratings and lending relationship

Variables	*(1)* *Banks ESG change*	*(2)* *Better borrower*	*(3)* *Worse borrower*
ESG_Diff $_{t-1}$	**0.030***	**0.045****	−0.010
	(0.015)	**(0.022)**	(0.042)
Borrower_Chg $_{t-1,\ t+1}$	0.032**	0.048***	0.014
	(0.015)	(0.018)	(0.025)
ESG_Bank$_{t-1}$	−0.596***	−0.868***	−0.420***
	(0.027)	(0.042)	(0.051)
lnPackageAmt	−0.061***	−0.044*	−0.028
	(0.021)	(0.026)	(0.034)
Firm Size	0.012	0.017	0.015
	(0.017)	(0.021)	(0.029)
ROA	0.032	−0.000	0.106
	(0.190)	(0.256)	(0.441)
Leverage	0.071	−0.089	0.133
	(0.123)	(0.151)	(0.189)
Market-to-Book	−0.001	0.010	−0.016
	(0.018)	(0.021)	(0.035)
Constant	1.572***	1.886***	0.597**
	(0.155)	(0.181)	(0.287)
Observations	3,303	1,689	1,374
R-squared	0.633	0.657	0.610
Year FE	Yes	Yes	Yes
Industry FE	Yes	Yes	Yes
Cluster	Yes	Yes	Yes
Adjusted R-squared	0.629	0.650	0.599

This table reports the OLS regression of the change in the bank's ESG profile on the ex-ante difference between the borrower and the bank's ESG ratings. The change in the bank's ESG profile (*Banks ESG Change*) is defined as the difference between the borrower's MSCI IVA ratings over a two-year window, from one year before to one year after the package initiation date. The ex-ante difference between the borrower and bank's ESG ratings (*ESG_Diff*) is defined as the difference between the borrower and bank's MSCI rating measured one year before the package initiation. *Borrower_Chg* controls for the evolution in the borrower's ESG ratings over the same two-year window. *ESG_Bank*$_{t-1}$ is defined as the bank's MSCI ratings one year before the loan initiation to address the potential path dependence concern. Each regression controls for Fama–French industry and year fixed effects. We additional control for package size (*lnPackageAmt*) and borrower's characteristics including *Firm Size, ROA, Leverage, and Market-to-Book ratio.* *, **, *** Indicates statistical significance at 10%, 5%, and 1% respectively

better ESG performance have a stronger disciplinary effect on a borrower's subsequent ESG changes. Similarly, we hypothesize that banks will benefit more when their borrowers have better ESG profiles. In other words, we expect our baseline result to be more pronounced among the subsample where borrowers have stronger ESG performance than their lending banks. To test our hypothesis, we split our sample into a "Better Borrower" group, where ESG_*Diff* > 0 and a "Worse Borrower" group, where (ESG_*Diff* < 0) based on the ESG gap between borrower and bank one year before the loan initiation. We rerun our regression model (2) within each group separately. Table 3, columns (2) and (3) show the subsample results. As expected, banks boost their ESG scores more if they are linked with relatively better ESG borrowers (column 2). It is noteworthy that, as shown in column 3, the coefficient of *ESG_Diff* is statistically insignificant, suggesting banks exposed to worse ESG borrowers are not motivated to improve their own ESG profiles. Our results complement the findings of Houston and Shan (2022) in showing that, in addition to the disciplining effect, banks also "learn by observing" the ESG activities of their borrowers, especially those of better ESG borrowers.

6 Conclusions

Over the past two decades, there have been growing concerns that environmental, social, and governance (ESG) issues could impose systemic risks upon the banking sector. In this chapter, we first explain the theory underlying why banks value their borrowers' ESG performance. We then review the recent studies that highlight how firms' ESG risks affect bank lending and investigate the real outcomes for borrowers via this bank lending channel. We also provide new evidence on the relationship between banks and ESG activities, specifically showing that bank ESG risks are value-relevant, since those risks provide incremental information to investors that changes how current earnings influence their views about the present value of future cash flows. Furthermore, such lending relationships help banks improve their ESG ratings by observing superior ESG borrowers' activities and learning from them. Overall, our study suggests that ESG is an important and relevant issue for banks whose responsibility it is to assist in socially responsible economic transformation.

Appendix 1: List of Published and Working Papers in ESG and Cost of Bank Loan

Primary variable	*Citation*
Environmental Factors	
Exposure to extreme winter weather	Brown et al. (2021)
Environmental performance	Chava (2014)
Environmental liability after the 2008 Apex Oil settlement	Chen et al., 2022
Exposure to climate change-related disaster: natural hazard events, including hurricanes, floods and wildfires	Correa et al. (2022)
Self-reported carbon emissions	Degryse et al. (2022)
Levels of fossil-fuel reserves	Delis et al., (2020a, 2020b)
Carbon emission intensity	Ehlers et al. (2022)
The California SB 32 on Greenhouse gas (GHG) emissions	Ivanov et al. (2021)
Exposure to climate-risks: drought intensity	Javadi and Masum (2021)
Firm-level carbon emissions and bank-level commitments to carbon neutrality	Kacperczyk and Peydró (2021)
Exposure to sea-level rise (SLR)	Nguyen et al. (2021)
Exposure to natural disaster: flooding	Ouazad and Kahn (2021)
Social Factors	
Employee treatment	Francis et al. (2019)
Social capital	Hasan et al., (2017)
Employee treatment	Qian et al. (2021)
Governance Factors	
Institutional ownership holding	Bhojraj and Sengupta (2003)
Takeover defenses	Chava and Roberts (2008)
ESG disclosure	Eliwa et al. (2021)
Board independence	Francis et al. (2012)
Female CFO	Francis et al. (2013)
Corporate governance index	Ge et al. (2012)
Female board member	Karavitis et al. (2021)
Internal control weaknesses disclosure	Kim et al. (2011)

Appendix 2: Variable Definitions

See Tables A1 and A2.

Table A1 The value-relevance of bank ESG risks

Variable	*Definition*
Dependent Variable	
Stock price	The stock price of firm *i* is the 3 months following the end of fiscal year *t*
Independent Variables	
EPS	The earnings per share is calculated as income before extraordinary items (*IB*) divided by the total number of shares outstanding at the end of fiscal year *t*
BVPS	The book value per share calculated as book value of equity (*CEQ*) divided by total number of shares outstanding (*CSHO*) at the end of fiscal year *t* following Clarkson et al. (2004)
RepRisk Index	The annual average of monthly *Current RRI* for firm *i* in year *t*. The *RepRisk Index* ranges from 0 to 100. The higher value, the higher ESG risks for firm *i* in year *t*
RepRisk Rating (RRR)	The annual average of monthly *RepRisk Rating* for firm *i* in year *t*. The *RepRisk Rating (RRR)* ranges from 0 to 10. The higher value, the higher ESG risks for firm *i* in year *t*
Control Variables	
Assets	The total assets (*AT*) scaled by the total number of shares outstanding at the end of fiscal year *t*
OCF	The operating cash flow (*OANCF*) scaled by the total number of shares outstanding at the end of fiscal year *t*
Cash	The cash and short investment (*CHE*) scaled by the total number of shares outstanding at the end of fiscal year *t*
Dividends	The dividends on common shares (*DVC*) scaled by the total number of shares outstanding at the end of fiscal year *t*
R&D	The R&D expense (*XRD*) scaled by the total number of shares outstanding at the end of fiscal year *t*. The missing R&D expense is replaced with zero
Intangibles	The intangible assets (*INTAN*) scaled by the total number of shares outstanding at the end of fiscal year *t*
Advertising	The advertising expense (*XAD*) scaled by the total number of shares outstanding at the end of fiscal year *t*. The missing advertising expense is replaced with zero

(continued)

Table A1 (continued)

Variable	*Definition*
Special Items	The Special Items (*SPI*) scaled by the total number of shares outstanding at the end of fiscal year *t*
OCI	The other comprehensive income (*OCI*) minus the prior year retained earnings (*RE*), plus the dividends on common shares (*DVC*) and minus the income before extraordinary items (*IB*) scaled by the total number of shares outstanding at the end of fiscal year *t*
Revenues	The total revenue (*REVT*) scaled by the total number of shares outstanding at the end of fiscal year *t*

Table A2 Banks ESG and lending relationships

Variable	*Definition*
Dependent Variable	
ESG_Chg_Bank	The change in bank's MSCI ESG ratings over the two-year window, before and after the loan initiation
Independent Variables	
ESG_Diff	The ex-ante difference between the borrower and bank's MSCI ESG ratings measured one year before the package initiation
Control Variables	
Borrower_Chg	The change in borrower's MSCI ESG ratings over the two-year window, before and after the loan initiation
ESG_Bank	The bank's MSCI ESG ratings one year before the loan initiation
lnPackageAmt	The logarithm of 1 plus package amount in million at the year *t*
Firm Size	The logarithm of 1 plus borrower's total assets (*AT*) at the year *t*
ROA	The net income (*NI*) scaled by the total assets (*AT*) at the end of fiscal year *t*
Leverage	The sum of borrower's long-term debt (*DLTT*) and short-term debt (*DLC*) scaled by total assets (*AT*) at the end of fiscal year *t*
MTB	The market value of equity (*CSHO* * *PRCC_F*) scaled by the book value of total assets (*AT*) at the end of fiscal year *t*

References

Adams, R. B., & Ferreira, D. (2009). Women in the boardroom and their impact on governance and performance. *Journal of Financial Economics, 94*(2), 291–309.

Ahn, S., & Choi, W. (2009). The role of bank monitoring in corporate governance: Evidence from borrowers' earnings management behavior. *Journal of Banking & Finance, 33*(2), 425–434.

Albuquerque, R., Koskinen, Y., & Zhang, C. (2019). Corporate social responsibility and firm risk: Theory and empirical evidence. *Management Science, 65*(10), 4451–4469.

Amiraslani, H., Lins, K. V., Servaes, H., & Tamayo, A. (2021). Trust, social capital, and the bond market benefits of ESG performance. *Review of Accounting Studies* (Forthcoming).

Avramov, D., Cheng, S., Lioui, A., & Tarelli, A. (2022). Sustainable investing with ESG rating uncertainty. *Journal of Financial Economics, 145*(2), 642–664.

Bae, K. H., & Goyal, V. K. (2009). Creditor rights, enforcement, and bank loans. *Journal of Finance, 64*(2), 823–860.

Balachandran, S., & Mohanram, P. (2011). Is the decline in the value relevance of accounting driven by increased conservatism? *Review of Accounting Studies, 16*(2), 272–301.

Barigozzi, F., & Tedeschi, P. (2015). Credit markets with ethical banks and motivated borrowers. *Review of Finance, 19*(3), 1281–1313.

Barth, M. E., Li, K., & McClure, C. G. (2022). *Evolution in value relevance of accounting information* (Working Paper).

Barth, M. E., Beaver, W. H., & Landsman, W. R. (1998). Relative valuation roles of equity book value and net income as a function of financial health. *Journal of Accounting and Economics, 25*, 1–34.

Barth, M. E., Beaver, W. H., & Landsman, W. R. (2001). The relevance of the value relevance literature for financial accounting standard setting: Another view. *Journal of Accounting and Economics*, *31*(1–3), 77–104.

Basu, S., Vitanza, J., Wang, W., & Zhu, X. R. (2022). Walking the walk? bank esg disclosures and home mortgage lending. *Review of Accounting Studies Forthcoming*.

Beaver, W. H. (1968). The information content of annual earnings announcements. *Journal of Accounting Research*, *6*, 67–92.

Becchetti, L., Ciciretti, R., Hasan, I., & Kobeissi, N. (2012). Corporate social responsibility and shareholder's value. *Journal of Business Research, 65*(11), 1628–1635.

Becchetti, L., & Manfredonia, S. (2022). Media, reputational risk, and bank loan contracting. *Journal of Financial Stability, 60*, 100990.

Beck, T., Demirgüç-Kunt, A., & Maksimovic, V. (2008). Financing patterns around the world: Are small firms different? *Journal of Financial Economics, 89*(3), 467–487.

Bénabou, R., & Tirole, J. (2010). Individual and corporate social responsibility. *Economica, 77*(305), 1–19.

Berg, F., Koelbel, J. F., & Rigobon, R. (2019). *Aggregate confusion: The divergence of ESG ratings* (pp. 1–42). Cambridge, MA, USA: MIT Sloan School of Management.

Beyene, W., Delis, M. D., De Greiff, K., & Ongena, S. (2020). Fuelling fossil fuel: Bond to bank substitution in the transition to a low-carbon economy, principles for responsible investment.

Bharath, S. T., Pasquariello, P., & Wu, G. (2009). Does asymmetric information drive capital structure decisions? *The Review of Financial Studies, 22*(8), 3211–3243.

Bharath, S. T., Dahiya, S., Saunders, A., & Srinivasan, A. (2011). Lending relationships and loan contract terms. *The Review of Financial Studies, 24*(4), 1141–1203.

Bhojraj, S., & Sengupta, P. (2003). Effect of corporate governance on bond ratings and yields: The role of institutional investors and outside directors. *The Journal of Business, 76*(3), 455–475.

Brown, J. R., Gustafson, M. T., & Ivanov, I. T. (2021). Weathering cash flow shocks. *The Journal of Finance, 76*(4), 1731–1772.

Cao, J., Liang, H., & Zhan, X. (2019). Peer effects of corporate social responsibility. *Management Science, 65*(12), 5487–5503.

Carney, M. (2015). *Breaking the tragedy of the horizon—Climate change and financial stability. [Speech of the Governor of the Bank of England at Lloyd's of London on 29 September].*

Chava, S. (2014). Environmental externalities and cost of capital. *Management Science, 60*(9), 2223–2247.

Chava, S., & Roberts, M. R. (2008). How does financing impact investment? The role of debt covenants. *Journal of Finance, 63*(5), 2085–2121.

Chava, S., Livdan, D., & Purnanandam, A. (2009). Do shareholder rights affect the cost of bank loans? *The Review of Financial Studies, 22*(8), 2973–3004.

Chen, J., Hsieh, P. F., Hsu, P. H., & Levine, R. (2022). *Environmental liabilities, creditors, and corporate pollution: Evidence from the Apex Oil ruling* (No. w29740). National Bureau of Economic Research.

Chih, H. L., Chih, H. H., & Chen, T. Y. (2010). On the determinants of corporate social responsibility: International evidence on the financial industry. *Journal of Business Ethics, 93*(1), 115–135.

Collins, D. W., Maydew, E. L., & Weiss, I. S. (1997). Changes in the value-relevance of earnings and book values over the past forty years. *Journal of Accounting and Economics, 24*(1), 39–67.

Core, J. E., Holthausen, R. W., & Larcker, D. F. (1999). Corporate governance, chief executive officer compensation, and firm performance. *Journal of Financial Economics, 51*(3), 371–406.

Correa, R., He, A., Herpfer, C., & Lel, U. (2022). *The rising tide lifts some interest rates: Climate change, natural disasters and loan pricing*. Available at SSRN: https://ssrn.com/abstract=3710451

Dai, R., Liang, H., & Ng, L. (2021). Socially responsible corporate customers. *Journal of Financial Economics, 142*(2), 598–626.

Dass, N., & Massa, M. (2011). The impact of a strong bank-firm relationship on the borrowing firm. *The Review of Financial Studies, 24*(4), 1204–1260.

Degryse, H., Goncharenko R,.Theunisz C., & Vadasz T. (2022). *When green meets green*. Available at SSRN 3724237.

Delis, M. D., De Greiff, K., & Ongena, S. (2020a). Being stranded with fossil fuel reserves? Climate policy risk and the pricing of bank loans. *Climate Policy Risk and the Pricing of Bank loans. EBRD Working Paper*, (231). Available at SSRN: https://ssrn.com/abstract=3451335

Delis, M. D., Hasan, I., & Ongena, S. (2020b). Democracy and credit. *Journal of Financial Economics, 136*(2), 571–596.

Dimson, E., Marsh, P., & Staunton, M. (2020). Divergent ESG ratings. *The Journal of Portfolio Management, 47*(1), 75–87.

Dyck, A., Lins, K. V., Roth, L., & Wagner, H. F. (2019). Do institutional investors drive corporate social responsibility? International evidence. *Journal of Financial Economics, 131*(3), 693–714.

Eccles, R. G., & Serafeim, G. (2013). *Corporate stewardship: Achieving sustainable effectiveness* (Ed. S. A. Mohrman, J. O' Toole, & E. E. Lawler III). Greenleaf Publishing, Routledge.

Edmans, A. (2011). Does the stock market fully value intangibles? Employee satisfaction and equity prices. *Journal of Financial Economics, 101*(3), 621–640.

Ehlers, T., Packer, F., & de Greiff, K. (2022). The pricing of carbon risk in syndicated loans: Which risks are priced and why? *Journal of Banking & Finance, 136*, 106180.

El Ghoul, S., Guedhami, O., Kwok, C. C., & Mishra, D. R. (2011). Does corporate social responsibility affect the cost of capital? *Journal of Banking & Finance, 35*(9), 2388–2406.

Fama, E. F. (1985). What's different about banks? *Journal of Monetary Economics, 15*(1), 29–39.

Fama, E. F., & Jensen, M. C. (1983). Separation of ownership and control. *The Journal of Law and Economics, 26*(2), 301–325.

Ferrell, A., Liang, H., & Renneboog, L. (2016). Socially responsible firms. *Journal of Financial Economics, 122*(3), 585–606.

Fitch Rating. (2020). *ESG in credit*. White paper.

Flammer, C. (2021). Corporate green bonds. *Journal of Financial Economics, 142*(2), 499–516.

Francis, B., Hasan, I., & Wu, Q. (2013). The benefits of conservative accounting to shareholders: Evidence from the financial crisis. *Accounting Horizons, 27*(2), 319–346.

Francis, B., Hasan, I., Koetter, M., & Wu, Q. (2012). Corporate boards and bank loan contracting. *Journal of Financial Research, 35*(4), 521–552.

Francis, B., Hasan, I., Liu, L., & Wang, H. (2019). Employee treatment and contracting with bank lenders: An instrumental approach for stakeholder management. *Journal of Business Ethics, 158*(4), 1029–1046.

Giglio, S., Kelly, B., & Stroebel, J. (2021). Climate finance. *Annual Review of Financial Economics, 13*, 15–36.

Goss, A., & Roberts, G. S. (2011). The impact of corporate social responsibility on the cost of bank loans. *Journal of Banking & Finance, 35*(7), 1794–1810.

Gul, F. A., Srinidhi, B., & Ng, A. C. (2011). Does board gender diversity improve the informativeness of stock prices? *Journal of Accounting and Economics, 51*(3), 314–338.

Ge, W., Kim, J.-B., & Song, B. Y. (2012). Internal governance, legal institutions and bank loan contracting around the world. *Journal of Corporate Finance, 18*(3), 413–432.

Hasan, I., Hoi, C. K. S., Wu, Q., & Zhang, H. (2014). Beauty is in the eye of the beholder: The effect of corporate tax avoidance on the cost of bank loans. *Journal of Financial Economics, 113*(1), 109–130.

Hasan, I., Hoi, C. K., Wu, Q., & Zhang, H. (2017). Social capital and debt contracting: Evidence from bank loans and public bonds. *Journal of Financial and Quantitative Analysis, 52*(3), 1017–1047.

Hong, H., & Kacperczyk, M. (2009). The price of sin: The effects of social norms on markets. *Journal of Financial Economics, 93*(1), 15–36.

Hong, H., Li, F. W., & Xu, J. (2019). Climate risks and market efficiency. *Journal of Econometrics, 208*(1), 265–281.

Houston, J. F., & Shan, H. (2019). Corporate ESG profiles and banking relationships. *The Review of Financial Studies*. Forthcoming

Houston, J., & James, C. (1996). Bank information monopolies and the mix of private and public debt claims. *The Journal of Finance, 51*(5), 1863–1889.

Houston, J. F., & Shan, H. (2022). Corporate ESG profiles and banking relationships. *The Review of Financial Studies*, *35*(7), 3373–3417.

Ilhan, E., Sautner, Z., & Vilkov, G. (2021). Carbon tail risk. *The Review of Financial Studies, 34*(3), 1540–1571.

Ilhan, E., Krueger, P., Sautner, Z., & Starks, L. T. (2021). Climate risk disclosure and institutional investors. *Swiss Finance Institute Research Paper*, 19–66.

Ioannou, I., & Serafeim, G. (2015). The impact of corporate social responsibility on investment recommendations: Analysts' perceptions and shifting institutional logics. *Strategic Management Journal, 36*(7), 1053–1081.

Ivanov, I., Kruttli, M. S., & Watugala, S. W. (2021). *Banking on carbon: Corporate lending and cap-and-trade policy*. Available at SSRN 3650447.

Ivashina, V. (2009). Asymmetric information effects on loan spreads. *Journal of Financial Economics, 92*(2), 300–319.

James, C. (1987). Some evidence on the uniqueness of bank loans. *Journal of Financial Economics, 19*(2), 217–235.

Javadi, S., & Masum, A. A. (2021). The impact of climate change on the cost of bank loans. *Journal of Corporate Finance, 69*, 102019.

Kacperczyk, M. T., & Peydró, J. L. (2021). *Carbon emissions and the bank-lending channel*. Available at SSRN 3974987.

Karavitis, P., Kokas, S., & Tsoukas, S. (2021). Gender board diversity and the cost of bank loans. *Journal of Corporate Finance, 71*, 101804.

Karpoff, J. M., Lott, J. R., Jr., & Wehrly, E. W. (2005). The reputational penalties for environmental violations: Empirical evidence. *The Journal of Law and Economics, 48*(2), 653–675.

Kim, J. B., Song, B. Y., & Zhang, L. (2011). Internal control weakness and bank loan contracting: Evidence from SOX Section 404. *The Accounting Review, 86*(4), 1157–1188.

Kim, S., Kumar, N., Lee, J., & Oh, J. (2022, March). *ESG lending.Proceedings of Paris December 2021 Finance Meeting EUROFIDAI* (Finance Working Paper No. 817/2022). ESSEC, European Corporate Governance Institute. Available at SSRN: https://ssrn.com/abstract=3865147

Klein, A. (2002). Audit committee, board of director characteristics, and earnings management. *Journal of Accounting and Economics, 33*(3), 375–400.

Kölbel, J. F., Busch, T., & Jancso, L. M. (2017). How media coverage of corporate social irresponsibility increases financial risk. *Strategic Management Journal, 38*(11), 2266–2284.

Krueger, P., Sautner, Z., & Starks, L. T. (2020). The importance of climate risks for institutional investors. *The Review of Financial Studies, 33*(3), 1067–1111.

Li, K., Liu, X., Mai, F., & Zhang, T. (2021). The role of corporate culture in bad times: Evidence from the COVID-19 pandemic. *Journal of Financial and Quantitative Analysis, 56*(7), 2545–2583.

Lins, K. V., Servaes, H., & Tamayo, A. (2017). Social capital, trust, and firm performance: The value of corporate social responsibility during the financial crisis. *the Journal of Finance, 72*(4), 1785–1824.

Lourenço, I. C., Callen, J. L., Branco, M. C., & Curto, J. D. (2014). The value relevance of reputation for sustainability leadership. *Journal of Business Ethics, 119*(1), 17–28.

Marquis, C., Toffel, M. W., & Zhou, Y. (2016). Scrutiny, norms, and selective disclosure: A global study of greenwashing. *Organization Science, 27*(2), 483–504.

Nguyen, D. D., Ongena, S., Qi, S., & Sila, V. (2021). Climate change risk and the cost of mortgage credit. *Swiss Finance Institute Research Paper* (20–97). Available at SSRN: https://ssrn.com/abstract=3738234

Ohlson, J. A. (1995). Earnings, book values, and dividends in equity valuation. *Contemporary Accounting Research, 11*(2), 661–687.

Ouazad, A., & Kahn, M. E. (2021). *Mortgage finance and climate change: Securitization dynamics in the aftermath of natural disasters* (No. w26322). National Bureau of Economic Research.

Qian, C., Crilly, D., Wang, K., & Wang, Z. (2021). Why do banks favor employee-friendly firms? *A Stakeholder-Screening Perspective. Organization Science, 32*(3), 605–624.

Raghunandan, A., & Rajgopal, S. (2021). *Do socially responsible firms walk the talk?* Available at SSRN: https://ssrn.com/abstract=3609056

Rajan, R., & Winton, A. (1995). Covenants and collateral as incentives to monitor. *The Journal of Finance, 50*(4), 1113–1146.

Reghezza, A., Altunbas, Y., Marques-Ibanez, D., d'Acri, C. R., & Spaggiari, M. (2021). *Do banks fuel climate change?* (ECB Working Paper No. 2021/2550). Available at SSRN: https://ssrn.com/abstract=3846654

Schwert, M. (2018). Bank capital and lending relationships. *The Journal of Finance, 73*(2), 787–830.

Securities and Exchange Commission (SEC). (2022). *SEC proposes rules to enhance and standardize climate-related disclosures for investors*. Available. https://www.sec.gov/news/press-release/2022-46.

Sen, S., & Bhattacharya, C. B. (2001). Does doing good always lead to doing better? Consumer reactions to corporate social responsibility. *Journal of Marketing Research, 38*(2), 225–243.

Servaes, H., & Tamayo, A. (2013). The impact of corporate social responsibility on firm value: The role of customer awareness. *Management Science, 59*(5), 1045–1061.

Shin, D. (2021). *Corporate esg profiles, matching, and the cost of bank loans* (Doctoral dissertation). University of Washington

Stern, N. (2007). *Stern Review: The economics of climate change*. Cambridge University Press.

Sufi, A. (2007). Information asymmetry and financing arrangements: Evidence from syndicated loans. *The Journal of Finance, 62*(2), 629–668.

Sufi, A. (2009). Bank lines of credit in corporate finance: An empirical analysis. *The Review of Financial Studies, 22*(3), 1057–1088.

Wu, M. W., & Shen, C. H. (2013). Corporate social responsibility in the banking industry: Motives and financial performance. *Journal of Banking & Finance*, *37*(9), 3529–3547.

ESG and Credit Risk

Chrysovalantis Gaganis, Fotios Pasiouras, and Menelaos Tasiou

1 Introduction

The acronym 'ESG', which stands for 'Environmental, Social and Governance', refers to a set of business metrics that socially conscious stakeholders use to observe—in quantifiable manners—a company's impact on society and sustainability. Whilst there are several sources of data on ESG key performance indicators (KPIs), one could envisage those dimensions as loosely capturing the impact of a firm on the *environment* ('E'—reducing carbon footprint, using renewable sources of energy and

C. Gaganis
Department of Economics, University of Crete, Rethymno, Greece
e-mail: c.gaganis@uoc.gr

F. Pasiouras (✉)
Montpellier Business School, Montpellier, France
e-mail: f.pasiouras@montpellier-bs.com

M. Tasiou
School of Accounting, Economics and Finance, University of Portsmouth, Portsmouth, UK
e-mail: menelaos.tasiou@port.ac.uk

C. Gaganis et al. (eds.), *Sustainable Finance and ESG*,
Palgrave Macmillan Studies in Banking and Financial Institutions,
https://doi.org/10.1007/978-3-031-24283-0_6

recycled products, developing or investing in green products/services), on the *society* ('S'—providing training and supporting health, safety and wellbeing, investing in the local community, preventing abuses within the supply chain and respecting human rights, ensuring customer rights and safety) and its capacity to have a sound *governance* in place ('G'—transparency in decision-making and mechanisms ensuring it, ethical processes and lack of bribery, diverse leadership teams and equality in pay) (see Inderst and Steweart, 2018, Appendix 2, for a broad view list of ESG criteria per pillar).

ESG issues are of interest to various stakeholders (Kay et al., 2020), and they have attracted a substantial attention in recent years by both academics and the general population (see Fig. 1). For example, the EY's Global Alternative Fund Survey report (EY, 2021, p. 6) states that: "[...] with environmental and social justice issues dominating the news cycle, the ESG movement has taken center stage with managers formalizing their ESG policies at both the management company level and in their investment strategies to satisfy investor demands". This demand on ESG investing may be explained by a variety of factors, such as better returns, more sound investing, minimization of downside risk related to the environment or reputation, or due to investment policy statement mandates (Natixis, 2021). Indeed, ESG criteria surface in investors' decision-making processes nowadays, with more than 50% of institutional investors including ESG characteristics in their decision-making processes during a screening process (52% considering 'diversity, equity and inclusion' attributes, 55% considering 'governance' attributes and 79% considering 'climate risk', highlighting the 'E' role of this responsibility umbrella term) (EY, 2021). A more thought-provoking survey by PwC finds that 76% of investors consider a company's exposure to ESG risks and opportunities when screening for potential investment opportunities; however, only 54% of investors agree that board directors are sufficiently knowledgeable about the ESG issues their companies face (PwC, 2021), highlighting a gap between investors' demand for more to be done on this front, and the lack of effort or knowledge in corporate boards to meet it.

In that regard, policymakers, regulators and investors are increasingly putting pressure on corporations by requiring disclosure and incorporation of ESG-related attributes in their financing decisions, in order to ensure sustainability and a sound financial system. A key reason is that lack of corporate responsibility comes at a premium when it comes to

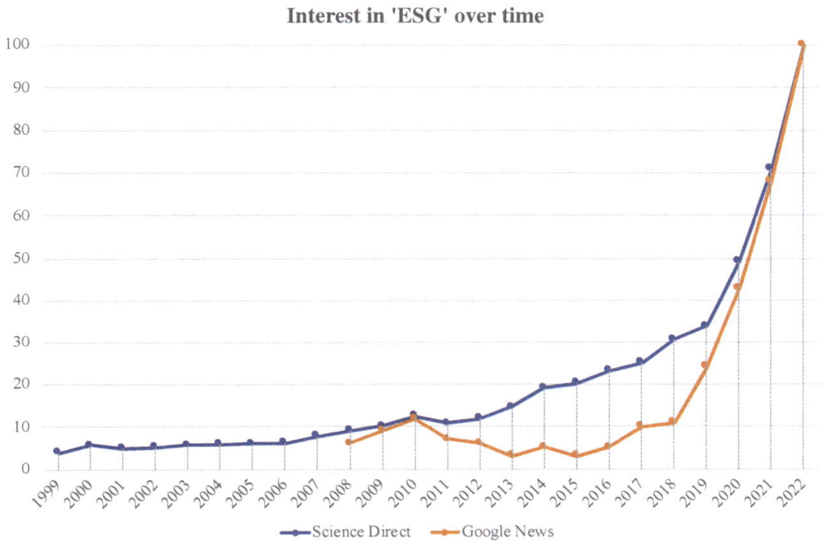

Fig. 1 Interest evolution in the term 'ESG' in academic studies and press. *Source* Authors' elaboration on data from SciencedDirect.com (published articles) and Google Trends (News searches). Illustrated data processed through the 'max' normalization technique for comparability

credit risk, which brings the relationship between ESG and credit risk into the spotlight, due to the long-term benefits it brings, but also due to the associated challenges that arise. This nexus is not new, as we are going to discuss in Sect. 2. However, integration of ESG factors into credit risk assessment is the most novel challenge for the financial industry in recent years (Brogi, 2020), and simultaneously a means to a more sustainable way forward.

In this chapter, we will explore the ESG-credit risk nexus with a twofold objective in mind. First, in Sect. 2 we discuss academic studies in the field, providing empirical evidence of this nexus and the implied mechanisms behind it. Second, in Sect. 3 we discuss how the three major rating agencies view and incorporate ESG aspects into their credit ratings.

2 ESG and Credit Risk: Academic Evidence

One of the earliest studies exploring the link between ESG and credit risk indicates that a large portion of credit losses in German banks could be retraced to environmental risks (Scholz et al., 1995, as cited in Henisz & McGlinch, 2019). Indeed, one may see how financial institutions may be (in)directly affected through environmental (legal or regulatory) risks associated with their investment decisions or reputation (Weber et al., 2010), which is why financial institutions discussed their commitment to the environment agenda a long time ago. Taking the United Nations Environment Programme (UNEP) as an example, a main result of its bank working party has been the development of a "Statement by Banks on the Environment and Sustainable Development" back in 1992 (UNEP, 1992), signed by 88 large banking institutions at the time, and subsequently altered to include more banking institutions after 1997. This has been the starting point for many similar initiatives and guidelines developed thereafter (Coulson & Monks, 1999), all in support of a more sustainable path for both the environment and the financial system.

Looking at a survey sample of European Banks, Weber et al. (2008) find that most institutions consider these risks in their credit appraisal process; however, not equally among all stages of the credit risk management process (rating, costing, pricing, monitoring and workout), with banks that signed up to the UNEP's finance initiative being more likely to consider the environmental risks in the credit appraisal process. Moreover, the integration of the social dimension into the credit model seems to improve the predictive validity of the credit rating process in the banks' loan portfolios (Weber et al., 2010). Furthermore, Goss and Roberts (2011) find support of a potential ESG-related rating process in banking institutions over a US sample. Their results show that below average ESG-performing firms typically pay 7 to 18 basis points premium on their bank loans, compared to their more "responsible" counterparts. Looking at the cost of equity capital side, Ghoul et al. (2011) find that the market discounts the cash flows of firms with good environmental performance, as more environmentally responsible firms are associated with a lower cost of equity capital. Why does this matter for credit risk? Increased capital costs are of significant interest to investors. As Bauer and Hann (2010, p. 4) mention, "firms that are held responsible for environmental violations can incur substantial clean-up costs, fines, and damage awards. [...]"

and indeed, less socially responsible firms face significantly more pollution and regulatory compliance violations than their responsible counterparts (Chatterji et al., 2009). It is therefore reasonable to assume that "[...] an inadequate management of environmental risks increases the likelihood of costly future liabilities, [as] bondholders should expect a higher default risk for borrowers with poor environmental practices, and a consequently impaired value of their investments" (Bauer & Hann, 2010, p. 4).

Of course, increased likelihood of default risk due to environmental practices is not the only link between ESG and credit risk. Attig et al. (2013) hypothesize that the mechanism between firm responsibility and its credit rating standing is the reduction of the firms' perceived risk of financial distress, running through three potential channels: (i) the improvement of relations with firms' stakeholders; (ii) the signalling of firms' efficient use of internal resources and robust performance; and (iii) the reduction of potential incurred costs associated with irresponsible firm behaviour. The arguments behind this threefold channel can be summarized by strategic management initiatives such as the "good management hypothesis" (Waddock & Graves, 1997). Changing expectations of stakeholders are pressuring strategic management to move from solely profit-maximizing objectives and include ESG-related attributes in their agenda. This could result in the improvement of a firm's reputation, therefore its relationship with its stakeholders and, eventually, the creation of intangible assets (customer loyalty, employee retention). These are essential to a firm's sustainability and, in turn, financial performance (Surroca et al., 2010). In the long run, this effect implies better profitability, reputation, lower cost of borrowing and, ultimately, reduced credit risk.

Still on the note of the underlying mechanism, it has been argued that disclosure of socially-oriented information—particularly related to pollution control—affects the perception of the public regarding the firm's level of compliance and the expectations about its future expected cash flows (Shane & Spicer, 1983). This is in line with the findings of El Ghoul et al. (2011), showing that investors perceive less responsible firms as having a higher idiosyncratic risk than their more responsible counterparts. Benlemlih et al. (2018) attribute this to stakeholder theory and the resource-based view of the firm. In particular, they suggest that companies with extensive and objective disclosures related to environmental and social responsibility promote corporate transparency which helps develop a positive reputation and trust with their stakeholders and, in turn, mitigate their idiosyncratic/operational risk.

Attig et al. (2013) hypothesize that credit analysts view ESG-related activities in their ratings decisions and provide further support for the link between environmental and social issues and credit standing. In more detail, their findings show that responsibility concerns are associated with a higher cost of debt financing and lower credit ratings. Similarly, the study of Capasso et al.'s (2020) indicates that the exposure to climate risks affects the creditworthiness of loans and bonds issued by corporates, as firms with higher carbon footprint are perceived by the market as more likely to default. Likewise, Oikonomou et al. (2014) find that social responsibility is favourably perceived by the market when it comes to the pricing of corporate debt and the assessment of credit quality of specific bond issues. Examining a range of credit risk measures, Bannier et al. (2022) find a negative association between firm responsibility traits related to the environment and society and their credit default risk. Using an international sample of more than 3000 firms from 79 countries, Brogi et al. (2022) conclude that higher ESG awareness is strongly associated with better creditworthiness (proxied by Altman's Z-score and the probability of default). Hsu and Cheng (2015) examine whether socially responsible firms behave differently from their counterparts when it comes to financial risk. Looking at a sample of US firms for over a 20 years period, the authors find that more responsible firms are associated with higher credit ratings and have lower credit risk (measured by bond spreads and distance to default). From an alternative methodological perspective, the study of Chang et al. (2013) estimates short-run and forward default probabilities for Taiwanese firms, linking those to responsibility attributes. The authors find that more responsible firms exhibit lower short-term and forward default probabilities.

In an effort to price ESG factors in credit markets, Reznik et al. (2021) introduced the "ESG-risk curve", a non-linear relationship between firms' ESG performance and their CDS spreads, which also holds in the case of the firms' credit ratings. The authors find that this link holds for both the overall ESG score and its individual dimensions. Looking at the relationship between ESG and CDS spreads, they find that firms in the fifth (top) quintile of ESG performance have, on average, a CDS spread of 82 basis points, compared to almost 300 for their counterpart in the first quintile of ESG performance. This is more than a threefold spread difference that increases in a non-linear manner. Interestingly, although the negative

CDS-ESG relationship is found to be true when looking at the individual pillars of ESG scores, the drop in basis points as the quintile of performance increases (curve slope) is more pronounced (steeper) for the 'E' and 'S' pillar, compared to the 'G' one. Thus, these two components of the composite umbrella term may potentially matter more when it comes to corporate credit risk measured by bond spreads. This comes as no surprise, and it is in accordance with the EY's (2021) survey, mentioning that "the most important driver of investor interest in sustainable funds is the environmental impact they could have". Höck et al. (2020) provide further evidence on the relationship between environmental sustainability and CDS. Using a sample of 149 European companies and data from the period 2006–2017. They conclude that more sustainable companies have lower credit risk premiums if they also have a high creditworthiness. To give an example, their estimates show that a company that has a CDS spread of 122.9 bp—which is the average CDS premium in their sample—can decrease its CDS spread by 2.1 bp by increasing its environmental score by 1 point.

Whilst the above studies provide consistent findings of the ESG-credit risk hypothesis, the study of Henisz & McGlinch, (2019) attempts to identify the specific material risks highlighted by poor ESG performance. The authors delve into a qualitative analysis that provides a variety of case studies where companies with relatively poor ESG performance experience losses after serious ESG-related incidents that led to significant jumps in their credit risk. The authors then proceed into finding a more generalized link of the ESG-credit risk nexus, over a sample of global projects involving similar material events from 13 industries. They conclude that the management of ESG-related risks is a capability that over the long term should correlate with fewer negative surprises—"fewer accidents, fewer lawsuits from aggrieved stakeholders, less government intervention into management practice, and fewer negative surprises on revenue, sales, and profitability" (Henisz & McGlinch, 2019, p. 117). It does not come as a surprise that qualitative information in the form of news or disclosure affects credit risk evaluation (Tsai et al., 2016), but it does provide evidence to otherwise conjectures that ESG-related incidents materialize into credit risk indeed through reputational exposure. What is more, this may be more pertinent to consider from an informal institution viewpoint, given that more environmentally friendly public perceptions are linked to lower reputational exposure in corporates (Gaganis et al., 2021).

All the above findings are further supported and materialized into improved practice through the "ESG in credit ratings" initiative of the UNEP and the Principles for Responsible Investment (PRI), which is supported by more than 180 institutional investors (with over US$40trn in collective AUM). Interestingly, the statement reads as follows: "We, the undersigned, recognise that environmental, social and governance (ESG) factors can affect borrowers' cash flows and the likelihood that they will default on their debt obligations. ESG factors are therefore important elements in assessing the creditworthiness of borrowers" (UNPRI, 2021), which perfectly aligns with the empirical findings of the aforementioned studies. Despite some opaqueness as for the exact ESG factors and the extent to which they are being considered, the initiative goes a long way in forging a more formal link in the ESG-credit assessment nexus, with the number of undersigned increasing since its first conception in 2017. That same year, the Task Force on Climate-related Financial Disclosure (TCFD) created by the Financial Stability Board (FSB), released climate-related financial disclosure recommendations designed to help companies provide better information to support informed capital allocation. TCFD (2017) recommended firms to disclose the metrics and targets they use to assess and manage relevant climate-related risks and opportunities where such information is material. According to a survey by the PRI, 65% of respondents claimed it has changed the way they conduct credit risk analysis (PRI, 2022).

Arguably, integrating ESG factors into credit risk assessment is gradually becoming the new frontier for credit risk management, and it is highlighted by empirical evidence and initiatives by established organizations, undersigned by financial institutions and investors globally. It is a significant challenge for corporate managers, but one they should seriously consider in their agendas given the stakeholders' ever-increasing demand for it. Therefore, it is not surprising that credit rating agencies (CRAs) have started incorporating such factors in their analysis. For example, Kiesel and Lücke (2019) use a latent dirichlet allocation (LDA) approach to examine ESG considerations in 3719 Moody's credit rating reports for USA and European listed firms between 2004 and 2015. They find evidence of a small but present consideration of ESG in rating decisions, corporate governance playing the most important role. Furthermore, a recent study by European Securities and Market Authority (ESMA) uses natural language processing techniques to analyse a unique dataset of over 64,000 CRA press releases published between 1

January 2019 and 30 December 2020, a recent study by the shows that the overall level of ESG disclosures in CRAs' press releases has increased since the introduction of the related ESMA guidelines in March 2020 (ESMA, 2022). In the following section, we focus on the methodology of the CRAs and discuss how they perceive ESG along with their role in fostering or facilitating these initiatives further when it comes to the impact of ESG on the assessment of creditworthiness.

3 ESG and Rating Agencies

In recent years, one after the other, the major rating agencies have paid particular attention to ESG issues and their incorporation into their assessments. Twenty seven CRAs have signed up to UNEP and PRI's "ESG in credit ratings" initiative, whilst all major rating agencies claim to incorporate ESG considerations in their analysis (S&P Global Ratings, 2017; Moody's 2017; FitchRatings, 2017). A pressuring reason is the demand for ESG-related information, given that 68% of investors use ESG ratings and scores when screening for potential investments (PwC, 2021). Despite empirical findings of the ESG-credit risk nexus, CRAs operate on opaque rating frameworks that are largely based on qualitative assessments of analysts based on a variety of factors. It is thus not always clear to what extent CRAs use and incorporate ESG factors in credit ratings, with some investors asking for a clarification of the role of ESG factors in ratings, or an explicit integration of ESG by CRAs (PRI, 2017).

In this section, we look at how the three major CRAs incorporate ESG-related information in their rating decisions, their common or different procedures to ratings' adjustment, as well as the transparency in reporting the effect running from ESG risk factors to issuers' creditworthiness.

The Case of Standard & Poor's (S&P)

S&P mentions the recognition of the increasing demand for ESG in credit ratings, and present the concept of 'ESG Credit Factors', the intersection of ESG factors and traditional credit factors that affect the current—or may influence the future—creditworthiness of an issuer. They state that "ESG credit factors are those ESG factors that can materially influence the creditworthiness of a rated entity or issue and for which we have sufficient visibility and certainty to include in our credit rating analysis.

ESG credit factors can have a negative or positive impact on creditworthiness, depending on whether they represent a risk or an opportunity, for example, regulatory or reputational risks" (S&P 2021a, p. 2).

S&P analyses ESG factors in two separate ways: (i) as part of their analysis on credit ratings and (ii) as part of their ESG-specific evaluations and opinions. The latter is a forward-looking assessment of an entity's ESG impact on broader stakeholders, including its relative performance and ability to prepare for future risks and opportunities. Within this context, S&P analyses how an entity is exposed to ESG issues along its value chain and its ability to manage through future disruption. As discussed in S&P (2021a), the ESG evaluation is neither a credit rating, nor a measure of credit risk, nor a component of their credit rating methodology. Therefore, we do not discuss it further and we rather focus on the incorporation of ESG factors into the credit ratings.

The issue of materiality is central in the way that S&P incorporates ESG factors in its analysis. As discussed in S&P (2021a), ESG factors typically incorporate an entity's effect on and impact from the natural and social environment and the quality of its governance; however, not all ESG factors materially influence creditworthiness and, consequently, credit ratings. Therefore, S&P defines ESG credit factors as those ESG factors that can materially influence the creditworthiness of a rated entity or issue and for which they have sufficient visibility and certainty to include in their credit rating analysis.

In their view, when they are sufficiently material to affect perceptions about creditworthiness, ESG credit factors can influence credit ratings through, for example: (i) a change in the size and relative stability of an obligor's current or projected revenue base, (ii) operating costs and requirements, (iii) risk planning, (iv) governance controls and standards, (v) profitability or earnings, (vi) cash flows or liquidity or (vii) the size and maturity of its financial commitments.

S&P provides the following examples of key ESG credit factors that have affected creditworthiness in the past or may influence it in the future: (i) Environmental Factors: climate transition risks; physical risks; natural capital; waste and pollution; other environmental factors, (ii) Social Factors: health and safety; social capital; human capital; other social factors, (iii) Governance Factors: governance structure; risk management, culture and oversight; transparency and reporting; other governance factors. Below are some examples of the potential influence of ESG credit factors on the corporate analysis, taken from (S&P, 2021b):

- **Climate transition risk**: Higher carbon dioxide emission costs leading to weaker profitability (reflected in the competitive position category of the corporate criteria) and debt service coverage ratios (cash flow leverage analysis)
- **Waste and pollution**: Fines imposed due to breach of pollution regulations leading to weaker profitability and liquidity
- **Health and safety**: Entities that suffer a drop in demand and revenues because of social distancing rules, including travel restrictions to stop the spread of virus, resulting in lower profitability
- **Social capital:** Aging population trends in advanced economies leading to sustainable positive growth in certain sectors (such as old age homes and health care and pharmaceutical companies), which is reflected in industry risk
- **Risk management, culture and oversight:** A history of regulatory, tax or legal infractions beyond an isolated episode or outside industry norms, creating liability risk that can affect a company's balance sheet (as part of the cash flow leverage analysis) or liquidity

S&P (2021b) also provides some potential examples in the case of financial institutions (banks and non-banks).

- **Climate transition risk**: A financial institution's (FI) risk position—which is one of the sector-specific factors in the financial institutions criteria—may be affected in cases that S&P anticipates the FI will suffer material charges due to the impact of climate transition risk on its loan and investment portfolios.
- **Physical risk:** Business position could come under pressure because of weakening asset quality amid more extreme climate conditions.
- **Social capital:** Lending activities that may be socially sensitive, such as high interest payday loans, can lead to reputation and regulatory risk (which we consider in our business position assessment)
- **Risk management, culture and oversight:** Litigation due to weaknesses in governance, risk appetite or the control framework leading to new risks not related to the credit quality of loans and investments, including, for example, money laundering or cyber risk (reflected in business position)

Alongside the description of ESG factors and how they may materialize in credit ratings, S&P provides more information about this link in the form of five general principles (S&P, 2021b). In particular, according to the first principle, the long-term issuer credit rating does not have a predetermined horizon, but it rather depends on how clear the view is about the factor potentially evolving and affecting an entity. An example of this, the CRA mentions, would be "an unexpected, drastic change in technology or customer behavior or extreme climate or political events that, while plausible, we may not have a view regarding their timing or likelihood". Hence, the CRA would typically monitor the factor's situation but not necessarily adjust a credit rating.

A second principle is that the influence of ESG factors on creditworthiness may largely vary by industry, geography and entity. For instance, exposure to climate transition risks may affect certain industries or firms in certain countries to a larger extent, whilst firm differences in how governance manages exposure risks, all of which can make a difference in how rating changes are affected.

A third principle relates to how the direction of and visibility into ESG credit factors may be uncertain and change rapidly. Factors may be more uncertain to foresee in the distant future compared to the short term. For instance, structural breaks in modelling may occur at certain points that are difficult to foresee in advance, due to, e.g. public policy decisions, changing public perceptions or even sudden extreme exogenous physical risks.

Tied to this principle, their fourth principle states that the influence of ESG credit factors may change over time if events crucial to forward-looking view of creditworthiness occur. They claim that "in some cases, a risk or strength that we currently consider immaterial to creditworthiness can later become material" (S&P, 2021a, p. 7).

Finally, their fifth principle simply states that a strong entity's creditworthiness should not necessarily correlate with strong ESG characteristics and vice versa. That is simply to say that ESG factors may evolve into ESG credit factors and affect ratings if they are material to the creditworthiness. They mention several examples, such as firms that may not necessarily have exceptional ESG attributes, but nevertheless have very sound financial fundamentals and pose no financial risk due to those ESG attributes not directly affecting—in a material way—the creditworthiness.

The Case of Fitch

Fitch developed an alternative system to show investors how ESG factors may affect a rating decision. In 2019, the CRA introduced the 'ESG Relevance Scores', an overall or pillar score rated on a scale between 1 to 5, illustrating how likely a particular factor is to materially affect a rating decision. Their scores are assigned by the analysts who rate an entity, with a score of '1' meaning a dimension is irrelevant to the entity rating and to the sector, and a score of '5' being highly relevant and a key rating driver significantly impacting the rating of an entity. According to the CRA, "the scores provide granularity on why ratings change and make the impact of ESG risks on a rating decision under Fitch's criteria much more transparent". The agency claims to be the first to have developed such an integration system due to "client demand and the large number of studies showing links between ESG factors and investing and financial performance" (FitchRatings, 2021, p. 8).

Similarly, to the case of S&P discussed before, Fitch claims that scores do not make value judgments on whether the engagement of a rated entity's good ESG practices, but the CRA focuses on the materiality of individual factors upon the creditworthiness of an entity. They note that ESG performance may align with credit risk, but this is not always necessary. An example of this, the CRA mentions, is carbon intensity (emissions per unit of revenue or energy produced), which is frequently used as an indicator of environmental performance. "While carbon intensity in itself is not relevant to credit analysis, it could be in jurisdictions where tighter regulation leads to additional costs associated with higher carbon intensity, or when changing social preferences present challenging financing conditions for carbon-intensive entities. The relevance to a credit rating will also depend on the broader credit profile, including the entity's ability to absorb or pass on higher costs, or its reliance on particular funding sources" (FitchRatings, 2021, p. 10).

According to the CRA's white paper (FitchRating, 2021), asset classes (corporates, financial institutions, sovereigns, public finance and infrastructure) range as regards how ESG factors are elevated and impact credit scores, with an average of 16% across asset classes presenting at least one elevated score impacting credit rating assessment. In the same paper, they mention that 'G' plays a role of paramount importance, followed by social and environmental factors, accordingly, which explains how the 'G' term in the responsible umbrella term goes a long way in explaining mitigation of potentially materializing events. However, Fitch mentions that as regulation intensifies and so does social pressure, this weight may radically change.

As in the case of S&P's, Fitch recognizes that ESG factors alternatively affect credit ratings in different sectors. The CRA discloses in detail how each sector's entity may be materially affected in each pillar according to the different asset class it belongs in. For instance, Fitch discloses 5 environmental (GHG emissions and air quality; energy management; water and wastewater management; waste and hazardous materials management; exposure to environmental impacts), 5 social (human rights, community relations, access & affordability; customer welfare; labour relations and practices; employee wellbeing; exposure to social impacts) and 4 governance (management strategy; governance structure; group structure; financial transparency) factors and how they may affect or not each sector, with the latter dimension's sub-pillars slightly changing according to the asset class.

The Case of Moody's

Understandably at this point, there are various similarities between all CRAs. Like the other two CRAs examined before, Moody's notes that ESG performance is not to be mistaken as a predictor of credit ratings. "Our objective is not to capture all considerations that may be labelled green, sustainable or ethical, but rather those that have a material impact on credit quality" (Moody's, 2017, p. 3). In addition, it notes that ESG factors may often be linked more to potential credit risk than credit benefit in otherwise lifting credit assessments: "As an example, a company with a track record of health and safety violations may face litigation risks that pressure its operating income, whereas another company that demonstrates outstanding health and safety practices may not see a comparable credit benefit". However, benefits could occur, e.g. "a company or government that has outstandingly strong governance is more likely to have a management culture of 360-degree risk assessment and informed decision-making, which support long-term creditworthiness" (Moody's, 2021, p. 3). Additionally, Moody's notes that assessments could largely vary per sector or geographic location, and the time span of ESG potential material effects on credit assessment is also reliant on 'visibility', i.e.

uncertainty about longer-term potential impacts is not necessarily incorporated in current credit ratings if there is a lot of uncertainty around it involved, and it could thus change in the future.

The CRA has a scorecard interface similar to that of Fitch, integrating ESG-related potential material events on the credit score of an entity on the basis of a 1–5 scale. Whilst the criteria may slightly differ according to the asset class, Moody's presents a general sub-pillar categorization of ESG-related credit factors per pillar (Moody's, 2021, p. 7). Environmental-related attributes include carbon transition, physical climate risks, water management, waste and pollution and natural capital issues. Social attributes include customer relations, human capital, demographic and societal trends, health and safety, responsible production issues. Governance attributes include financial strategy and risk management, management credibility and track record, organizational structure, compliance and reporting and board structure and policies issues.

Moody's presents some general principles about the assessment of environmental, social and governance risks. Starting with the first, the CRA sees environmental risk falling under two types: (i) consequences of regulatory or policy initiatives that seek to reduce or prevent environmental trends or hazards or perceived trends or hazards; (ii) adverse effects of direct environmental trends and hazards, such as pollution, drought, severe natural and human-caused disasters and climate change. Turning to social risk, Moody's differentiates between public and private issuers, with the former's attributes related to developing, executing and adjudicating laws, regulations and policies that address the needs of society, and the latter's attributes related to considerations about product safety, supply-chain considerations, business reputation or employee relations. Reasonably, according to the CRA, social issues may take years to emerge as credit concerns in the public sector, in contrast to the private sector. Finally, turning to the 'G' risk assessment, when it comes to the public sector, Moody's mainly considers the quality of institutions, broadly classifying the risk in four categories: (i) institutional structure; (ii) policy credibility and effectiveness; (iii) transparency and disclosure; and (iv) budget management. As regards the private sector issuers' assessment, the CRA considers risks in five categories, namely: (i) financial strategy and risk management; (ii) management credibility and track record; (iii) organizational structure; (iv) compliance and reporting; and (v) board structure, policies and procedures.

4 Conclusions

Interest in ESG has been increasing exponentially across a variety of stakeholders. Its link to credit risk is not new but has been signified in recent times. Findings consistently support the negative relationship between the two, which, alongside initiatives that bolstered ESG's place in credit monitoring and analysis, brought this nexus into the spotlight. In this chapter, we have explored academic and market research focusing on this nexus, uncovering the underlying mechanisms behind it and highlighted its materiality. Despite a variety of studies documenting this effect, surveys from institutional investors show that there is still a lot of progress to be made on many fronts but, ultimately, entities and rating agencies need to catch up with investor's demands and facilitate more sustainable capital allocation. The former can do more on the sustainability front, the latter need to be more transparent in their rating decisions and frameworks.

In that regard, we have also explored the CRAs frameworks when it comes to the assessment of creditworthiness and how this is affected by ESG factors. Although responsible initiatives have been initiated long ago, recent initiatives by established bodies immensely helped with progress in the incorporation of ESG factors into credit risk assessment. Evidently, CRAs have many common frameworks in how they approach this nexus, and they have been more transparent in their creditworthiness assessment modelling. However, lack of standardized ESG frameworks, lack of disclosure and opaqueness in qualitative credit assessment decisions means there is still room for improvement.

References

Attig, N., El Ghoul, S., Guedhami, O., & Suh, J. (2013). Corporate social responsibility and credit ratings. *Journal of Business Ethics*, *117*, 679–694.

Bannier, C. E., Bofinger, Y., & Rock, B. (2022). Corporate social responsibility and credit risk. *Finance Research Letters, 44*, 102052.

Bauer, R., & Hann, D. (2010). Corporate environmental management and credit risk. *Available at SSRN 1660470.*

Benlemlih, M., Shaukat, A., Qiu, Y., & Trojanowski, G. (2018). Environmental and social disclosures and firm risk. *Journal of Business Ethics, 152*(3), 613–626.

Brogi, M. (2020). ESG data. In *Joint JRC-EBA workshop on banking regulation and sustainability* (p. 32).

Brogi, M., Lagasio, V., & Porretta, P. (2022). Be good to be wise: Environmental social, and governance awareness as a potential credit risk mitigation factor. *Journal of International Financial Management & Accounting, 33*, 522–547.

Capasso, G., Gianfrate, G., & Spinelli, M. (2020). Climate change and credit risk. *Journal of Cleaner Production, 266*, 121634.

Chang, T. C., Yan, Y. C., & Chou, L. C. (2013). Is default probability associated with corporate social responsibility? *Asia-Pacific Journal of Accounting & Economics, 20*(4), 457–472.

Chatterji, A. K., Levine, D. I., & Toffel, M. W. (2009). How well do social ratings actually measure corporate social responsibility? *Journal of Economics & Management Strategy, 18*(1), 125–169.

Coulson, A. B., & Monks, V. (1999). Corporate environmental performance considerations within bank lending decisions. *Eco-Management and Auditing: The Journal of Corporate Environmental Management, 6*(1), 1–10.

El Ghoul, S., Guedhami, O., Kwok, C. C., & Mishra, D. R. (2011). Does corporate social responsibility affect the cost of capital?. *Journal of banking & finance, 35*(9), 2388–2406.

ESMA. (2022, February 10). *Text mining ESG disclosures in rating agency press releases, ESMA 80–195–1352.* Available at: https://www.esma.europa.eu/sites/default/files/library/esma80-195-1352_cra_esg_disclosures.pdf

EY. (2021). *Can the difference of one year move you years ahead?* 2021 Global Alternative Fund Survey. Available at: https://www.ey.com/en_us/news/2021/11/ey-global-alternative-fund-survey

FitchRatings. (2017). *FitchRatings approach to capturing environmental, social and governance risk in credit ratings.*

FitchRatings. (2021). *ESG in credit*. White Paper. Available: https://www.fitchratings.com/research/corporate-finance/esg-in-credit-white-paper-2021-20-07-2021

Gaganis, C., Papadimitri, P., Pasiouras, F., & Ventouri, A. (2021). Informal institutions and corporate reputational exposure: The role of public environmental perceptions. *British Journal of Management, 32*(4), 1027–1061.

Goss, A., & Roberts, G. S. (2011). The impact of corporate social responsibility on the cost of bank loans. *Journal of banking & finance, 35*(7), 1794–1810.

Henisz, W. J., & McGlinch, J. (2019). ESG, material credit events, and credit risk. *Journal of Applied Corporate Finance, 31*(2), 105–117.

Höck A., Klein C., Landau A., & Zwergel B. (2020). The effect of environmental sustainability on credit risk. *Journal of Asset Management, 21*, 85–93

Hsu, F. J., & Chen, Y. C. (2015). *Is a firm's financial risk associated with corporate social responsibility?. Management decision.* https://doi.org/10.1108/MD-02-2015-0047

Inderst, G., & Stewart, F. (2018, April). *Incorporating environmental, social and governance (ESG) factors into fixed income investment*. World Bank Group Publication.

Kay, I., Brindisi, M., & Martin, B. (2020). *The stakeholder model and ESG*. In Harvard Law School Forum. Available: https://corpgov.law.harvard.edu/2020/09/14/the-stakeholder-model-and-esg/

Kiesel, F., & Lücke, F. (2019). ESG in credit ratings and the impact on financial markets. *Financial Markets, Institutions & Instruments, 28*, 263–290.

Moody's. (2017). *Moody's approach to assess ESG in credit analysis*.

Moody's. (2021). *General principles for assessing environmental, social and governance risks methodology*. Available at: https://esg.moodys.io/esg-credit

Natixis. (2021). *2022 Institutional Outlook*. Available at: https://www.im.natixis.com/intl/resources/2022-institutional-outlook-executive-overview

Oikonomou, I., Brooks, C., & Pavelin, S. (2014). The effects of corporate social performance on the cost of corporate debt and credit ratings. *Financial Review, 49*(1), 49–75.

PRI. (2017). *Shifting perceptions: ESG, credit risk and ratings. Part 1: The state of play*.

PRI. (2021). *Statement on ESG in credit risk and ratings*. Available: https://www.unpri.org/credit-risk-and-ratings/statement-on-esg-in-credit-risk-and-ratings-available-in-different-languages/77.article

PRI. (2022). *Are the TCFD recommendations useful for credit risk analysis?*. Available at: https://www.unpri.org/credit-risk-and-ratings/are-the-tcfd-recommendations-useful-for-credit-risk-analysis/9429.article

PwC. (2021, December). *PwC's global investor survey. The economic realities of ESG*. Available at: https://www.pwc.de/de/nachhaltigkeit/pwc-global-investor-survey-2021.pdf

Reznik, M., Viehs, M., Panesar, T. (2021). *Pricing ESG risk in credit markets*. Federated Hermes International. Available at: https://www.hermes-investment.com/uploads/2021/03/3c2c00030f63fa43992b459ec164497a/fhi-fixed-income-research-pricing-esg-in-credit-markets-03-21.pdf

S&P Global Ratings. (2017). *How does S&P global ratings incorporate environmental, social, and governance risks into its ratings analysis*.

S&P Global Ratings. (2021a, November 17). *ESG credit factors. A deeper dive*. Available at: https://www.spglobal.com/_assets/documents/ratings/research/100701190.pdf

S&P Global Ratings. (2021b, October 10). *Environmental, social, and governance principles in credit ratings*.

Scholz, R. W., Weber, O., Stünzi, J., Ohlenroth, W., & Reuter, A. (1994). *Umweltrisiken systematisch erfassen/Systematically inventorying environmental risks* (Vol. 10). Schweizer Bank: Monatsmagazin für Führungskräfte aus Bank und Finanz/Swiss Bank: Magazine for Managers in Banks and Finance.

Shane, P. B., & Spicer, B. H. (1983). Market response to environmental information produced outside the firm. *Accounting Review, 58*, 521–538.

Surroca, J., Tribó, J. A., & Waddock, S. (2010). Corporate responsibility and financial performance: The role of intangible resources. *Strategic Management Journal, 31*(5), 463–490.

Task force on climate-related financial disclosures. (TCFD) (2017). *Recommendations of the task force on climate-related financial disclosures*. Available: https://assets.bbhub.io/company/sites/60/2021/10/FINAL-2017-TCFD-Report.pdf

Tsai, F. T., Lu, H. M., & Hung, M. W. (2016). The impact of news articles and corporate disclosure on credit risk valuation. *Journal of Banking & Finance, 68*, 100–116.

UNEP. (1992). Banking and the environment: A statement by banks on the environment and sustainable development. United Nations Environment Programme.

UNPRI. (2021). *Statement on ESG in credit risk and ratings*. Available at: https://www.unpri.org/credit-risk-andratings/statement-on-esg-in-credit-risk-and-ratings-available-in-different-languages/77.article

Waddock, S. A., & Graves, S. B. (1997). The corporate social performance–financial performance link. *Strategic Management Journal, 18*(4), 303–319.

Weber, O., Fenchel, M., & Scholz, R. W. (2008). Empirical analysis of the integration of environmental risks into the credit risk management process of European banks. *Business Strategy and the Environment, 17*(3), 149–159.

Weber, O., Scholz, R. W., & Michalik, G. (2010). Incorporating sustainability criteria into credit risk management. *Business Strategy and the Environment, 19*(1), 39–50.

The Politics of Climate Finance and Policy Initiatives to Promote Sustainable Finance and Address ESG Issues

Paola D'Orazio

1 Introduction

The world is confronted with several social, environmental, and economic issues. Poverty reduction, climate change mitigation, economic inequality reduction, and, more recently, pandemic risks necessitate substantial financial resources and investments. Sustainable development and green recovery are shared goals pursued by countries worldwide to reconcile the conflict between economic growth and environmental preservation.

The United Nations' Sustainable Development Goals (SDGs) and the Paris Agreement have established the international community's commitment to a more sustainable society and a climate-neutral economy. A new technological framework and a shift in consumption patterns were highlighted—among others—as essential and interrelated aspects that would steer the transformation in order to attain these ambitious goals.

P. D'Orazio (✉)
Chair of Macroeconomics, Ruhr-Universität Bochum, Bochum, North Rhine-Westphalia, Germany
e-mail: paola.dorazio@rub.de

C. Gaganis et al. (eds.), *Sustainable Finance and ESG*,
Palgrave Macmillan Studies in Banking and Financial Institutions,
https://doi.org/10.1007/978-3-031-24283-0_7

However, significant environmental improvement is impossible without sufficient financial support; therefore, policymakers and researchers agree that so-called green finance is critical in aiding the transition to a low-carbon and climate-resilient economy (Roy et al., 2013; Sachs et al., 2019).

Article 2.1(c) of the Paris Agreement commits financial flows to "a path-way toward low greenhouse gas emissions and climate-resilient development" (COP, 2015). Nevertheless, numerous studies have found that there is a significant financial gap in meeting these objectives (see Buchner et al., 2017b; IPCC, 2018, among others). According to GCoA (2019), $1.8 trillion in potential investments would yield $7.1 trillion in benefits, while Climate Policy Initiative estimates $30 billion in adaptation investments (Buchner et al., 2017a). Despite significant advances, green financing faces various obstacles that affect the supply and demand for green products. As noted by Bhandary et al. (2021), some of these challenges are specific to green projects; others are more generic and apply to most long-term endeavors.[1]

In light of the fast-paced evolution of events and global discussions, this chapter examines the steps at the national and international levels and the financial sector's strategies and tools for scaling up green finance and coping with climate-related financial risks. The proposed analysis identifies gaps and challenges in existing policies and policy frameworks and analyzes global experiences and prospective future research and policies.

Regarding disclosure requirements, the Task Force on Climate-related Financial Disclosures has recently set a benchmark for financial and corporate disclosures and reporting climate-related financial information. Additionally, risk disclosure has been a subject of initiatives to create greater investor openness. The rationale is that investors should be informed about the environmental impact of their investments, and financial institutions should be transparent about how they assess environmental risk (Alessi et al., 2021; Fiedler et al., 2021).

[1] They cite a lack of quantitative incentives, most for-profit firms' inability to absorb environmental externalities, poor, or intangible returns to corporate social responsibility efforts, commercial banks and other mainstream perceptions of high risks of low-carbon technology, a mismatch between long-term payback periods and the short-term horizons of most private investors, a lack of information to evaluate projects, and a lack of information to evaluate projects as examples.

Concerning green taxonomy regulation, the review finds that it has been introduced across several jurisdictions, spearheaded by the European Union. However, a lot remains to be done in this area. Regarding green financial instruments, the Network for Greening the Financial System has contributed to the international alignment of sustainable monetary policy and incorporating climate change into risk assessments in the past years. However, the spread of these practices varies by country, and there are significant gaps in existing frameworks and jurisdictions regarding mitigating climate risks and scaling up green finance (D'Orazio, 2022; D'Orazio & Popoyan, 2019). Governments and multilateral organizations have taken several steps to promote sustainable financing in response to the Paris Agreement. Following a so-called Paris effect, they have emphasized voluntary activities, relying on businesses and investors to self-report on this topic. Nevertheless, while many financial institutions have already implemented internal systems for assessing and monitoring their businesses' sustainability, regulatory frameworks focusing on sustainable finance are still lagging behind (D'Orazio, 2021).

The investigation carried out in the chapter highlights that the lack of adequate financial resources to be devoted to the transition to a sustainable economy could be understood and explained by considering the current state of policies adopted at the international and national levels aimed at affecting the financial sector. The non-standardized and non-mandatory disclosure requirements of ESG (Environmental, Social, and Governance) factors and risks, the lack of internationally agreed-upon taxonomies of economic activities, and the lack of financial instruments to scale up green finance and/or address climate risks are the major aspects being considered in this framework. These measures are beginning to be implemented into legislation as the global transition to a more sustainable economy accelerates, and the need to address the challenges posed by climate change intensifies (D'Orazio, 2022). The rationale is that increased knowledge of climate risks and transparency enhance climate risk assessment and capital allocation to sustainable investments.

The chapter is organized as follows. Section 2 introduces the debate on sustainable finance and the main issues in green financial policymaking. Section 3 briefly illustrates the method used in the investigation. Section 4 presents the analysis of the major international initiatives aimed at fostering green financial markets and addressing ESG issues, and reports the survey results to analyze the diffusion of policies at the national level. Section 5 discusses the implications of the findings of the study and Sect. 6 provides concluding remarks.

2 Background

Sustainable finance refers to financial firms incorporating environmental, social, and governance (ESG) aspects into their business or investment choices[2] (Jebe, 2019). It considers social factors such as working conditions, local communities, conflict, and human rights, as well as governance issues such as executive pay, bribery and corruption, board structure, and tax strategy, in addition to environmental issues such as reducing environmental impact, protecting natural capital, minimizing waste, and reducing greenhouse gas emissions (Cunha et al., 2021; Fatemi & Fooladi, 2013).

Climate change is a major topic in sustainable finance. Climate finance (usually defined as "financing to fund actions that reduce greenhouse gas emissions or help adapt to climate change" under the Paris Agreement) has expanded, although—as explained above—sustainable finance has a broader scope and is linked to the UN Sustainable Development Goals. Environmental, social, and governance challenges can significantly impact the functioning of banks and the financial system. Environmental disasters have resulted in significant losses for banks and insurers (Batten et al., 2016; Kron et al., 2019) and social risks, such as inequality, may persuade policymakers to promote household borrowing for consumption, resulting in financial instability in the long run.

Existing literature identifies two main risk channels of transmission from climate change to the financial system (Carney, 2015). Changes in the price of stranded assets (such as coal and oil that will not be used during the fossil fuel phase-out) and economic disruptions caused by climate-related policies, technology, and market attitude during the transition to a lower-carbon economy are all sources of transition risks. Damages from weather-related events and broader climate changes are examples of physical risks.

Financial risks from climate change are difficult to calculate, but most research points to trillion-dollar economic and financial costs. Since the 1980s, insurance losses from climate-related natural disasters like droughts, floods, and wildfires have doubled, and asset values may not

[2] In particular, the acronym ESG frames the notion in terms of material risks posed by the environmental and social factors to businesses.

fully internalize climate risk and the shift to a cleaner economy (Caldecott et al., 2021). Therefore, a lack of awareness of these risks and delayed action to tackle them could jeopardize financial stability (D'Orazio & Popoyan, 2019; D'Orazio, 2021).

While scaling up climate finance remains a difficulty, there is widespread agreement that investments managed to consider broader sustainability criteria have increased steadily, and ESG integration into sustainable investing[3] has become more popular in recent years (Folqué et al., 2021; Maiti, 2021). Furthermore, according to the Global Sustainable Investment Alliance, sustainable investments in the major five developed economies increased by 34% in two years (between 2016 and 2018) (GSIA, 2019). Exclusion criteria and ESG integration were the most often used ESG approaches, accounting for over USD 37 trillion and two-thirds of all assessed sustainable investments (about 6% of investments when considered in the aggregate), with novel strategies including screening and sustainability-themed investing showing a substantial increase.

In recognition of the risks climate change could pose to financial and non-financial institutions, there is a growing call to disclose the risks they face from the physical impacts of climate change and the transition to a low-carbon economy. The idea is that disclosure will help them prepare for climate change impacts and—at the same time—help investors understand risks to make more informed investment decisions. Nevertheless, questions emerge about how companies should assess and report these risks, including whether the disclosures should be mandatory (see Fiedler et al., 2021, among others). Moreover, there is still much uncertainty when it comes to the financial sector's tangible impact on climate change and sustainable development (Kölbel et al., 2020), and there is a growing demand for a more thorough evaluation of ESG scores, including greater standardization of scoring systems and a shared understanding of the many ESG criteria and their substantial influence on climate change mitigation (BCBS, 2021; FSB, 2022; IPCC, 2021; Popescu et al., 2021).

[3] As an example, consider that in 2020, the UN Principles for Responsible Investment (PRI) had over 3000 signatories, representing more than USD 100 trillion in assets under management.

3 Method and Research Approach

The research approach used in this chapter is as follows. First, we set up the framework of the analysis, that is green finance and the policies aimed at fostering green markets, ESG factors disclosure requirements, green taxonomies, and financial measures. Second, we define the scope of the analysis: the focus is especially on G20 countries as they represent the major economies and are responsible for most CO2 emissions globally, denoting the most important players in achieving global adaptation and mitigation targets. Accordingly, the major national and international initiatives are considered, including actions promoted by international organizations and standard-setting bodies. Third, we define the time span to be considered. The investigation covers the period 2000 to 2021 to consider the most recent engagement at the country and international levels. Fourth, we define the search terms used in the survey. We consider keywords on green finance, the banking industry and financial regulation, including 'finance', 'financing', 'loan', 'credit', 'investment', 'banking', 'bank', 'financial institutions', 'banking sector', 'financial regulation', 'financial policies', 'promotional credit', 'prudential', and 'financial principles'. These keywords were combined with 'green', 'sustainable', 'climate-related', 'environmental', and 'sustainable' to restrict the search to the policies related to climate risks and low-carbon transition. Furthermore, information on (1) the authority in charge of the development or promotion of the policy and (2) the final "beneficiary" of the policy has been collected. Regarding the former, we distinguish between central banks, financial supervisors, and government or non-governmental actors. Regarding the latter, we distinguish between banks and non-financial institutions.

The findings of the review are presented in Sect. 4 and discussed in Sect. 5.

4 Analysis

International Initiatives

At the international level, the global engagement in climate finance started in the early 2000s, as shown in Fig. 1, and over the years, the financial industry has established many frameworks that are now part of what is known the "sustainable finance landscape" (Buchner et al., 2017a). Several actors, including the United Nations, the Financial

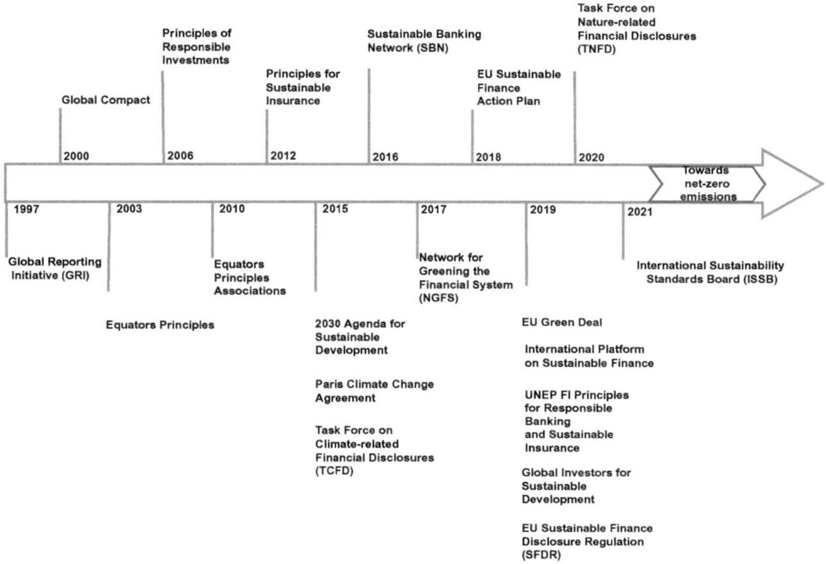

Fig. 1 Timeline of the adoption of international agreements aimed at fostering climate finance and promoting environmental, social, and governance factors (*Source* Author elaboration)

Stability Boards, the Sustainable Banking Networks, and the Network for Greening the Financial System—among others—are providing guidance to the financial sector in transitioning to a green economy.

The *Global Reporting Initiative* (GRI), which published its Sustainability Reporting Standards in 2015 following the approval of the United Nations' sustainable development objectives[4] was one of the first reporting frameworks to integrate ESG criteria. The Global Sustainability Standards Board is in charge of establishing sustainability reporting standards. These guidelines are intended to be economy-wide, comprising

[4] The GRI was founded in 1997 by the non-profit organization Ceres (previously the Coalition for Environmentally Responsible Economies) and Tellus Institute in the United States, supported by the United Nations Environment Programme (UNEP). Despite its independence, the GRI remains a UNEP collaborating center and collaborates with the United Nations Global Compact.

sector-specific, and universal standards, to provide a "single language" for ESG impact communication.

Other relevant initiatives followed—including the *Global Compact*[5] in 2000 and the *Equators Principles (EPs)*[6] - in 2003; a crucial turning point in the sustainable finance landscape adoption timeline was 2015, when multiple initiatives were implemented internationally.

2015 marks the adoption of the *2030 Agenda for Sustainable Development* by all UN member states. The Agenda consists of 17 Sustainable Development Goals (SDGs) and 169 associated targets aimed at ending global poverty and hunger, combating inequalities within and between countries, building peaceful, just, and inclusive societies, protecting human rights and promoting gender equality and women's empowerment, and ensuring the long-term protection of the planet and natural resources. In that year, the *Paris Agreement* was also signed. The Paris Agreement is a climate change international convention covering climate change mitigation, adaptation, and finance topics. It aims to improve parties' ability to respond to climate change effects and raise necessary funds. No mechanism obligates a government to set precise emissions goals, although each goal should be higher than the preceding one. As outlined by the Paris Climate Agreement and the United Nations Sustainable Development Goals (SDGs), nations have agreed to and are committed to ambitious global goals. Over 200 countries joined

[5] On January 31, 1999, then-UN Secretary-General Kofi Annan unveiled the UN Global Compact in a speech to the World Economic Forum, and it was formally launched on July 26, 2000. The UN General Assembly has designated the Global Compact Office as an entity that "promotes responsible business practices and UN principles throughout the global business sector and the UN System". Along with the Principles for Responsible Investment (PRI), the United Nations Environment Programme Finance Initiative (UNEP-FI), and the United Nations Conference on Trade and Development, the UN Global Compact is a founding member of the United Nations Sustainable Stock Exchanges (SSE) initiative.

[6] The EPs, formally launched in Washington, DC, on June 4, 2003, were built on the International Finance Corporation's existing environmental and social policy frameworks. The EPs are a risk management methodology used by financial institutions in project finance to determine, assess, and manage environmental and social risk. Its primary goal is to provide minimal due diligence to facilitate prudent risk decision-making. They have been officially endorsed by 116 financial institutions in 37 countries (as of March 2021), covering the majority of international Project Finance debt in emerging and established economies. The EPs apply globally to all industry sectors and four financial products, namely, (1) Advisory Services for Project Finance, (2) Funding for projects, (3) Corporate Loans for Projects, and (4) Bridge Loans.

the Agreement to improve the global response to climate change by limiting "global average temperature increases to below 2 degrees Celsius over pre-industrial levels". To attain this goal, a move to a low-carbon economy is required; nevertheless, this transformation may result in an ambiguous landscape of risks and opportunities from a market standpoint (Carney, 2015, 2019).

2015 also marks another important milestone. The Financial Stability Board (FSB) established the Task Force on *Climate-Related Financial Disclosures* (TCFD) to develop consistent climate-related financial risk disclosures for corporations, banks, and investors to deliver information to stakeholders in order for financial markets to price climate-related risks and opportunities accurately. The rationale is that increasing the amount of reliable data on financial institutions' exposure to climate-related risks and opportunities will improve financial system stability, contribute to a better understanding of climate hazards, and make financing the transition to a more stable and sustainable economy easier. The TCFD is one of the most prominent global standards of the last few years and is focused on developing data and projection tools and bridging the climate risk data gap. The financial and corporate sectors have adopted the TCFD as a mandatory disclosure tool in several jurisdictions, including the United Kingdom, Brazil, and Japan (TCFD, 2021).

More recently, recognizing the relevance of biodiversity loss for financial stability (NGFS, 2021), the *Task Force on Nature-related Financial Disclosures* (TFND) was created as a framework for organizations to report and act on nature-related risks based on the structure and basis of the TCFD (TNFD, 2022). To reduce repetition and maximize the chances of accelerated market adoption, it exploits commonalities in framework design and stakeholder interaction with the TCFD.

In 2016, it was established the *Sustainable Banking Network* (SBN); a knowledge-sharing and capacity-building forum comprising emerging market financial regulators, banking organizations, and environmental regulators dedicated to developing sustainable finance frameworks based on national context and priorities as well as international best practices (SBN, 2020), which is now known as the Sustainable Banking and Finance Network (SBFN). It is facilitated by the IFC and funded by the World Bank Group and assists members in mobilizing information, resources, and practical support to plan and implement national initiatives that advance sustainable finance at the national, regional, and global levels. The members are dedicated to transforming their financial sectors to be more sustainable, with the objectives of (i) improving the financial sector's management of environmental, social, and governance (ESG) risks, particularly climate hazards; (ii) increasing investment in initiatives that have positive environmental and social consequences, such as climate change mitigation and adaptation (SBFN, 2022).

Another important landmark in the international green finance framework is the creation of the *Network for Greening the Financial Sector* (NGFS) in 2017 (NGFS, 2017). The NGFS is not a standard-setter; rather, it is a voluntary network of central banks and supervisors and was formed to meet the Paris Agreement's goals and strengthen the financial system's role in risk management and capital mobilization for green and low-carbon investments in the context of environmentally sustainable development (NGFS, 2019). As of February 2022, the NGFS membership consists of 108 central bank members and 17 observers, representing 85 percent of global greenhouse gas emissions.

In 2019, the European Commission produced the *Sustainable Finance Disclosure Regulation* (SFDR), an obligatory disclosure instrument linked to the EU taxonomy. It is aimed toward players in the financial markets and asset managers that must provide standardized disclosures on ESG criteria and their integration at the entity and product level. The SFDR establishes sustainability disclosure requirements for financial product producers and financial advisers in dealings with investors. It refers to financial market participants and advisers, including sustainability risks in all investment processes and products with a long-term investment goal. At the entity and financial product levels, it also includes transparency obligations for negative implications on sustainability. For end investors, the goal is to make financial products and funds more comparable and transparent.

The EU is also drafting the *Corporate Sustainability Reporting Directive* within its green taxonomy, which will be released in October 2022. Participants will be obliged to report on how sustainability concerns affect their business and the impact of their actions on people and the environment. It is meant for corporations with listed and non-listed company requirements.

The actions taken at the European level are particularly interesting for the development of green finance and related regulations at the global level, and signals the willingness of the European Union of "leading by example". The European Commission has promoted an action plan, whose objectives[7] are as follows:

- To reorient capital flows toward sustainable investment to achieve sustainable and inclusive growth.
- To manage financial risks stemming from climate change, environmental Degradation, and social issues.
- To foster transparency and long-termism in financial and economic activities.

These efforts have received support and recognition from banking supervisors and regulators, who have demonstrated the importance of sustainability issues by forming task forces such as the Financial Stability Board's Task Force on Climate-related Financial Disclosures (2017, 2018).

More recently, the International Financial Reporting Standards (IFRS) established the *International Sustainability Standards Board* (ISSB) to address the development of diverse disclosure and reporting systems and assure their interoperability and standardization. The creation of the ISSB was announced at COP26 in November 2021 and represents a first step in developing a global, baseline corporate reporting standard on climate change and sustainability. Its goal is to provide investors and other capital market participants with information about a company's sustainability risks and possibilities so that they may make educated decisions.

National Actions

This section outlines the policies that have been adopted at the national level as a result of the above-mentioned international engagement. It would be expected that increased international engagement, as shown

[7] These objectives are—in turn—supported by ten actions, which include: (i) establishing an EU classification system for sustainable activities; (ii) creating standards and labels for green financial products; (iii) fostering investment in sustainable projects; (iv) incorporating sustainability when providing financial advice; (v) developing sustainability benchmarks; (vi) better-integrating sustainability in ratings and market research; (vii) clarifying institutional investors' and asset managers' duties; (viii) incorporating sustainability into prudential requirements; (ix) strengthening sustainability disclosure and accounting rule-making; and (x) fostering sustainable corporate governance and attenuating short-termism in capital markets.

in section "International Initiatives", would be matched by increased commitment at the national level. Indeed, evidence shows that the public sector has been more involved in integrating standards and policy efforts and promoting financial industry transparency and disclosure standards. However, our findings emphasize that present accomplishments are insufficient to address climate-related financial risks and foster green finance (D'Orazio, 2021); the public sector must take additional steps to improve its essential role in enabling a long-term transition.

Disclosure Requirements

The debate on climate risks prompted the development or discussion of quantitative exercises to measure the financial system's resilience to climate exposures in various circumstances (Monasterolo, 2020). Some G20 countries, such as the United Kingdom and France, have been engaged in green finance policymaking since the early 2000s, with the adoption of climate-related disclosure requirements (primarily for non-financial institutions, pension funds, and insurance companies) and so-called green finance principles and guidelines aimed at creating a financial market aligned with climate change concerns, according to our survey. Others, such as the German financial regulatory agency BaFin and the Banco Central do Brazil, have recently focused on climate risks, and are considering sustainability risk management and ESG disclosures (see D'Orazio, 2021, for a broader discussion on these topics).

The Climate Disclosure Standards Board has noted several implementation challenges, including a lack of internal and investor participation, a lack of board-level education, difficulty adapting to longer-term perspectives, and out-of-date risk management and financial modeling tools. As a result, there is a wide range of implementation (from no engagement in Russia and Saudi Arabia to encoding into law in France). Nevertheless, despite being a voluntary framework, the TCFD principles are intended to be widely adopted, useful to investors and lenders, and seek forward-looking financial impact information. According to TCFD (2021) only Brazil, the European Union, Japan, and the UK have adopted the recommendations among the G20 countries. Other countries extra-G20 reported to be active in this area are Hong Kong, Singapore, Switzerland, and New Zealand.

In the European Union, the High-level expert group on sustainable finance (HLEG) recommended that the current legislative framework, particularly the Non-Financial Reporting Directive 2014/95/EU, should be assessed for its suitability in light of the EU's Taxonomy Regulation.

Regarding the United States, instead, it strives to catch up with the EU on disclosure frameworks, as noted in OMFIF (2022).

Green Taxonomies

Concerning the green taxonomy, several experiences are detected internationally.

In the EU, the green taxonomy debate started in 2018, with the first proposal by the European Commission (EC, 2018), followed by the establishment of a Technical Expert Group (TEG) on sustainable finance in July 2018. The TEG published its final report in March 2020, providing recommendations on the design and guidance on how companies and financial institutions can use and disclose against the taxonomy (TEG, 2020). Because of political negotiations, the regulation entered into force only in July 2020.[8]

In some emerging economies, regulatory authorities have adopted green taxonomies starting in 2015 (OECD, 2020a). For example, in China, the Green Finance Committee of the China Society for Finance and Banking, a subsidiary of the People's Bank of China, issued a Green Bond Endorsed Project Catalogue (GFC, 2015). Additional green standards and classification methods have been published in China. Responsible authorities are ministries, commissions, and regional authorities, including the China Securities Regulatory Commission, the National Development and Reform Commission, and the China Banking and Insurance Regulatory Commission.

Other relevant experiences in Asia are provided by Bangladesh and Mongolia, which adopted green banking guidelines in 2017 and 2019, respectively (BB, 2020; FSCM, 2019).

Sustainable Finance Regulations in Financial Institutions

Financial institutions' reaction to the widespread concerns of sustainability issues and climate risks has been heterogeneous across countries (see D'Orazio, 2022, for a recent analysis).

[8] Among EU countries, we note that France was the first country to issue a sovereign green bond in 2017; nevertheless, it does not have a sustainable finance taxonomy per se (OECD, 2020b).

In recent years, the debate has centered on how central banks and regulators might assist in the transition to a more sustainable economy by including sustainability factors into their risk management models and governance framework (D'Orazio & Popoyan, 2022). Customers and investors, who demanded more low-carbon financial products, but also supervisors, who are paying greater attention to climate-related financial concerns, exerted pressure in this direction (D'Orazio & Popoyan, 2019; McInerney & Bunn, 2019; NGFS, 2020; OMFIF, 2020; Sachs et al., 2019).

Because no universally accepted "green taxonomy" exists yet (Steuer & Tröger, 2022), financial institutions have relied on various international frameworks and standards to identify ESG components, though some have developed and used their own definitions (Coletaon et al., 2020). There is, thus, currently a lack of commonality on how to address ESG factors (see OMFIF, 2022, for a recent review of the main issues). This raises concerns because if financial institutions adopt alternative definitions of ESG criteria, the risk management outputs and disclosure findings may differ, challenging the effectiveness of supervisors' role.

The European Banking Authority (EBA) has been mandated to promote sustainable finance models by the introduction and revisions of the EBA Regulation No 1093/2010, Capital Requirements Regulation No 575/2013, and Directive 2013/36/EU, Investment Firms Regulation 2019/2033 and Directive 2019/2034, and the European Commission's Action Plan on Sustainable Finance. Much of this work is planned to be completed by 2025. According to this framework, ESG risks will be incorporated into the oversight of institutions' risk management policies, national regulators' review processes, and stress-testing methodologies to uncover climate-related risks, exposures, and other vulnerabilities. For that purpose, the EBA released a discussion paper on the management and supervision of ESG risks for credit institutions and investment businesses in November 2020, outlining how ESG factors and ESG risks are identified and explained, focusing on environmental hazards, particularly climate change (EBA, 2020a).

Overview of the Survey Results

The survey of the policies included in this section is based on the taxonomy developed in D'Orazio (2022). In particular, the analysis considers the (i) so-called Green Financial Principles (GFP), which are defined as measures to promote the creation of green or climate-aligned financial markets, (ii) disclosure requirements for banks (DISCL-FI), and (iii) disclosure requirements aimed at non-financial institutions, such as, e.g., pension funds (DISCL-NFI).

We start by observing the total number of policies adopted, as shown in Fig. 2. Overall, the European Union and the UK are the most active in promoting GFP policies in this direction, having promoted and adopted 14% and 13% of the total measures, respectively. The adoption of disclosure requirements for banks is very low as we found only five policies at the country level. Engagement in disclosure requirements for non-financial institutions is more common, and the most active countries are Australia, China, France and United States. Considering the total number of policies adopted until 2021, we observe that the major adopters are high-income countries (68%), thus confirming a heterogeneous approach to greening the financial system (D'Orazio, 2022; D'Orazio & Popoyan, 2022).

The timing of the adoption is also relevant and interesting for our analysis. In Fig. 3, we report the cumulated value of adopted policies surveyed in our study. First, we observe that the increase in the trend started after 2015, in correspondence with the adoption of the Agenda for Sustainable Development, the Paris Agreement, and the creation of the Task Force on Climate-related Financial Disclosures. This confirms what is sometimes called as "Paris effect" in the literature, with a peak of adoptions recorded in both 2018 and 2019. Figure 3 shows that some activity was recorded at the beginning of the 2000s, but it was very sparse (and mostly promoted in high-income countries).

Regarding the type of authority responsible for promoting and implementing these policies, we find that central banks and financial regulators represent only 25%; the bulk is rather in the hands of nationally elected governments.

5 Discussion

Toward Mandatory and Internationally Harmonized Disclosure Requirements

The TCFD has played an important role in increasing climate finance in financial markets (including the private sector) by providing information to price climate risks and related opportunities. As discussed in the Section "National Actions", risk disclosures are crucial because they

Country	GFG	DISCL-FI	DISCL-NFI
Argentina	3	0	0
Australia	6	0	7
Brazil	4	0	4
Canada	2	0	5
China	6	2	7
EU	14	0	4
France	9	0	8
Germany	5	0	2
India	2	0	3
Indonesia	2	1	2
Italy	3	0	4
Japan	9	0	1
Mexico	4	1	0
Russian Federation	6	0	4
Saudi Arabia	0	0	0
South Africa	3	0	1
South Korea	2	0	1
Turkey	1	1	0
United Kingdom	13	0	4
United States of America	7	0	6
TOTAL	101	5	63

Fig. 2 Total adoption of green financial guidelines (GFG), Disclosure Requirement for Financial Institutions (FI) and Non-financial Institutions (NFI) in G20 countries in the period 2000 to 2021 (*Source* Author elaboration)

inform market participants about carbon-intensive asset concentrations at the portfolio level, allowing stakeholders to analyze banks' ESG risks and long-term financing strategies.

As reported in TCFD (2021), many of the existing disclosure frameworks are already mandatory, and in the European Union, financial institutions must report their investments and portfolio allocations under the SFDR. Nevertheless, the review's findings show that corporations do not report on sustainability regularly or consistently, particularly concerning the environmental and social dimensions. This makes it difficult for investors to incorporate ESG principles into their portfolios (OMFIF, 2022). Similarly, financial institutions identify ESG factors using a variety of international frameworks and standards, albeit some use their own definitions. This indicates the absence of uniformity in ESG considerations and raises concerns for the supervisory mandate because if

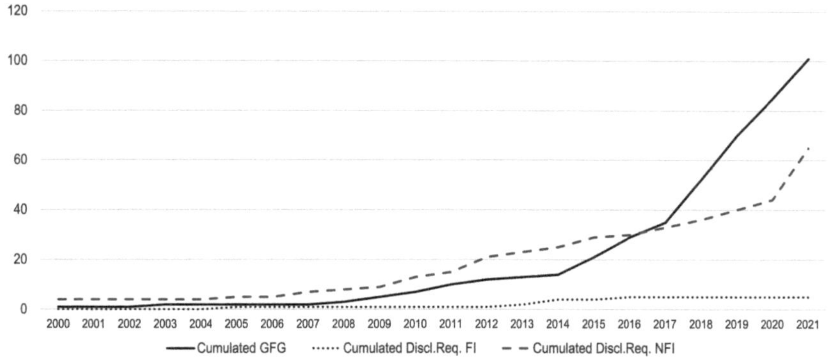

Fig. 3 Cumulated number of green financial guidelines (GFG), Disclosure Requirement for Financial Institutions (FI), and Non-financial Institutions (NFI) in G20 countries in the period 2000 to 2021 (*Source* Author elaboration)

financial institutions adopt alternative definitions of ESG criteria, the risk management outputs, and disclosure outcomes may differ (EBA, 2020b).

Moreover, it should be noted that mandatory disclosures may impose costs on smaller financial institutions and enterprises that lack the expertise or resources to report on their ESG compliance. Smaller financial institutions and organizations, in particular, may face costs as a result of obligatory disclosures since they lack the expertise and resources to report on their environmental, social, and governance (ESG) compliance.[9] Furthermore, different countries will be characterized by different conditions and capacities, making mandatory global disclosure difficult. On the one hand, financial institutions operating in emerging countries may find it difficult to comply with certain global rules or responsibilities due to a lack of data, inadequate supervisory capabilities, or government backing demanding such disclosures. On the other hand, requiring ESG disclosures may not be a pressing issue in some countries.

Obstacles continue despite great progress in adopting internationally agreed-upon disclosure and reporting standards. Disclosure of climate-related risk and investments and understanding how to assess impact

[9] Legal and regulatory costs, as well as the cost of data collecting and processing deriving from mandatory disclosure requirements, may worsen these issues by putting financial burdens on companies.

utilizing numerous data sources to benchmark costs is particularly complex. This is especially true when calculating risk exposure in the face of highly interconnected value chains (see recent discussion in FSB, 2022).

Finally, we argue that disclosure requirements may not be enough to solve the problems of climate finance alignment if they are not examined alongside other prudential regulations and made mandatory and consistent at the international level (Ameli et al., 2019, 2021).

Lack of Data to Understand Climate Risks and Growing Greenwashing Concerns

Data quality and availability are a huge challenge for the global financial industry to recognize climate risk, define targets, and drive ESG investments and products (BCBS, 2021). On the one hand, investors require reliable data from their investee companies to meet escalating regulatory obligations. Asset managers, central banks, multinational enterprises, and multilateral development banks use this data to conduct scenario analysis and stress testing to understand their climate risk exposure better. On the other hand, countries and jurisdictions do not have the same skills for accessing, gathering, aggregating, and verifying data, which is a big hurdle to mandated disclosure and reporting.

These issues must be addressed at the international level: forward-looking estimates and CO2 emissions data must be integrated into the financial sector's measurements and frameworks to assess and capture the risks associated with climate change accurately and adequately communicate the impact of investments and portfolios. As a result, demand for predictions and forward-looking statistics that include physical and transition risk is growing (see recent discussions in BoE-PRA, 2021; FSB, 2022).

Our survey revealed that quantifying the impact of ESG initiatives on attaining goals like reducing emissions is difficult (Gatti et al., 2019; Yu et al., 2020). For example, as noted in (OMFIF, 2022), investors, particularly public sector pension funds, find it difficult to implement ESG principles into their investments due to conflicting information on their performance and impact. Firms also confront challenges: while incorporating ESG elements into their business models has the potential to benefit them, the beneficial results are usually long-term, while the high costs of the disclosure are immediate (Uyar et al., 2020).

Greenwashing, or making false claims about an asset's or fund's ESG compliance, is also a risk that can harm a company's reputation (Khan, 2022; Popescu et al., 2021; Ruiz-Blanco et al., 2022). Greenwashing is a significant threat and is especially prevalent in new green asset classes (e.g., ESG funds, Green asset-backed securities or index-linked green products), while a clearer framework benefits the diffusion of green bonds (although the lack of controls and sanctions still poses a risk). Overall, the lack of rules, sanctions, and investors' lack of understanding to properly examine investment policies and environmental risks prevent the expansion of green financial markets. We argue that stricter rules and further international efforts are needed to overcome greenwashing and promote the development of green financial markets.

Climate-Aligned Financial Measures to Promote Green Finance

Because of pressure from customers, investors, and financial regulators, banks have begun to recognize sustainability risks, and support the transition to a more sustainable economy by incorporating sustainability factors into their risk management models and governance frameworks. Nevertheless, investigating the existing micro- and the macro-prudential frameworks, we find that, on the one hand, they are not aligned with the Paris Agreement goals. On the other hand, they do not handle crucial aspects such as cross-sectorial, global, and systemic dimensions. As a result, climate risks are not fully captured and are only indirectly reflected, and the potential green finance catalyst role is hampered.

Overall, the analysis and evidence suggest that monetary and prudential tools can facilitate the mitigation role of carbon pricing and affect CO2 emissions (Cœuré, 2018; D'Orazio & Dirks, 2021; Krogstrup & Oman, 2019), and they should be more actively promoted at the global level.

Capital regulations might be enacted to encourage financial institutions to allocate capital to green, carbon-neutral assets. However, modifying capital requirements to favor green assets or penalize polluting assets would offer major problems to policymakers and have a severe impact on the financial system's resilience without internationally harmonized taxonomies and disclosure methodologies. Therefore, as argued in previous sections, further efforts must be developed in these areas to allow more "audacious" policies aimed at promoting green finance and investments to be implemented.

Green Bonds and the Development of Green Financial Markets

Finally, before providing concluding remarks, we consider whether the policies examined in Section 4 have been beneficial in fostering the development of green markets in terms of green bonds diffusion.

The green bond[10] market is gaining momentum as a viable financial option for climate change mitigation (Braga et al., 2021; Chen & Zhao, 2021). According to Climate Bond Initiative, green bonds have gained popularity, with more than 290 billion dollars issued in 2020 and 1.1 trillion dollars in outstanding bonds (CBI, 2020, 2021). Corporations issued the highest volumes, financial institutions, and government-backed organizations (e.g., real estate, retail, manufacturing, and energy utilities), with proceeds pre-dominantly toward GHG reduction in energy, buildings, and transportation projects.

Studies have uncovered green bond pricing into the concept of a "green premium" (also sometimes known as "greenium") and insights into the important features and drivers that influence it (MacAskill et al., 2021). However, methodological differences among these studies have resulted in a general lack of agreement on the existence of the green premium. Overall, existing literature reveals a lack of understanding of green product structural properties. Many investors are unfamiliar with green goods' risk and return characteristics, which may be worsened by the lack of credit ratings and historical data. Some research highlights that green bonds are similar to standard bonds in terms of financial qualities. It is argued that, except for the ring-fencing of funds required by the green label, green bonds have the same financial characteristics as conventional bonds issued by the same issuers (e.g., same credit quality). Consequently, green bonds do not appear to have a pricing advantage over conventional bonds. Other studies show an agreement on the existence of a green premium, particularly for government-issued green bonds that are investment-grade and follow defined green bond governance and reporting protocols (MacAskill et al., 2021).

Nevertheless, besides the growing success, it is still unclear whether green bonds can act as a possible source of climate finance because of the insufficient amount of green investments to meet the needs of investors, as emphasized by existing literature (Banga, 2019; Bhutta et al., 2022).

[10] Green bonds are frequently referred to as climate bonds because they concentrate on GHG mitigation; however, the prevalent market nomenclature is "green".

The rationale is that green bonds, like conventional bonds, require many green assets to be issued, and certain corporations may have difficulty finding them to issue bonds. Moreover, green bonds may cause investors to be cautious because they perceive them as less liquid than other assets. As a result, liquidity is scarce, and many countries' green finance markets are still small and immature (Bhutta et al., 2022).

6 Conclusions

The international and national initiatives to scale up green financing over the last few decades have been studied in this chapter. In particular, it has examined climate finance politics and policy initiatives, such as the Paris Agreement and the EU Action Plan on Sustainable Finance, and academic evidence on the impact of these measures on market characteristics such as green bond market growth.

Several "policy areas", according to the analysis, still need to be improved, necessitating international cooperation and further action at the country level to accomplish a low-carbon transition. The following are the significant concerns and future directions uncovered in this investigation.

First, financial institutions usually base their disclosures on data from their investee companies, but the mechanism for disclosing and reporting, gathering ESG data, and setting transition targets are rarely standardized. To close the data gap, policymakers and regulators must cooperate in reconciling real-world and financial-world disclosure standards to help close the data gap. Financial and non-financial institutions must incorporate disclosure and reporting frameworks into their strategies.

Second, regarding climate risks, banks continue to under-disclose whether climate and environmental change have a meaningful impact on their risk profile and how the transition and physical risk affect their strategy. Climate change risks can be difficult to predict, depending on the rate of technical advancement (such as renewable energy deployment) and the level of future greenhouse gas emissions (with higher emissions, and higher temperatures, posing a greater threat to climate damages). Thus, financial institutions must use scenario analysis and stress-testing methods to incorporate climate risk into their strategy. Policymakers and regulators must help by establishing tools and frameworks for constructing and evaluating risk mitigation actions in the financial sector. Obliging them to consider climate risk would necessitate increased monitoring and enforcement by the appropriate authorities. However, measuring the influence of such policies on capital flows is difficult, and they may fall short of fostering a low-carbon transition if they are not accompanied by fiscal, monetary, and financial policies. Furthermore, capital regulations might be enacted to encourage financial institutions to allocate capital to green, carbon-neutral assets. Modifying capital requirements to favor green assets or penalize polluting assets would offer major problems to policymakers and have a severe impact on the financial system's resilience without internationally harmonized taxonomies and disclosure methodologies.

Third, amid concerns of greenwashing, reliable sustainability assessment methods are needed to ensure that funds are channeled toward "priority" sectors for the transition to a low-carbon and more inclusive economy.

Fourth, green assets are in short supply. Organizations preparing to originate green loans and/or issue green bonds frequently fail to describe and identify green assets on their balance sheet due to a lack of clear definitions and indicators to assess the 'greenness' of assets (i.e., standardized taxonomies at the international level). In addition, the existing data management systems are inadequate.

The research presented in this chapter has demonstrated the importance of harmonizing global sustainable finance norms, for which some crucial actions must be accomplished. Overall, the findings highlight the need for further research in these areas to promote green finance diffusion and support the low-carbon transition. In particular, we emphasize that a standardized and mandatory disclosure and reporting system will drive more data and ESG-related risk information; this will also enable the development of forward-looking projections and achievable low-carbon sustainable goals.

References

Alessi, L., Ossola, E., & Panzica, R. (2021). What Greenium matters in the stock market? The role of greenhouse gas emissions and environmental disclosures. *Journal of Financial Stability, 54*, 100869.

Ameli, N., Drummond, P., Bisaro, A., Grubb, M., & Chenet, H. (2019). Climate finance and disclosure for institutional investors: Why transparency is not enough. *Climatic Change, 160*, 1–25.

Ameli, N., Kothari, S., & Grubb, M. (2021). Misplaced expectations from climate disclosure initiatives. *Nature Climate Change, 11*(11), 917–924.

Banga, J. (2019). The green bond market: A potential source of climate finance for developing countries. *Journal of Sustainable Finance & Investment, 9*(1), 17–32.

Batten, S., Sowerbutts, R., & Tanaka, M. (2016). *Let's talk about the weather: The impact of climate change on central banks* (Tech. Rep.). Bank of England.

BB. (2020). *Sustainable finance policy for banks and financial instittutions* (Tech. Rep.). Bangladesh Bank—Sustainable Finance Department.

BCBS. (2021). *Climate-related financial risks—Measurement methodologies* (Tech. Rep.). BCBS—Basel Committee on Banking Supervision—Bank for International Settlements.

Bhandary, R. R., Gallagher, K. S., & Zhang, F. (2021). Climate finance policy in practice: A review of the evidence. *Climate Policy, 21*(4), 529–545.

Bhutta, U. S., Tariq, A., Farrukh, M., Raza, A., & Iqbal, M. K. (2022). Green bonds for sustainable development: Review of literature on development and impact of green bonds. *Technological Forecasting and Social Change, 175*, 121378.

BoE-PRA. (2021, October 28). Climate-related financial risk management and the role of capital requirements, climate change adaptation report 2021. *Bank of England—Prudential Regulation Authority.*

Braga, J. P., Semmler, W., & Grass, D. (2021). De-risking of green investments through a green bond market–empirics and a dynamic model. *Journal of Economic Dynamics and Control, 131*, 104201.

Buchner, B., Clark, A., Falconer, A., Macquarie, R., Meattle, C., Tolentino, R., & Cooper, W. (2017a). *Global landscape of climate finance 2019* (Tech. Rep.). Climate Policy Initiative.

Buchner, B., Oliver, P., Wang, X., Carswell, C., Meattle, C., & Mazza, F. (2017b). *Global landscape of climate finance 2017* (Tech. Rep.). Climate Policy Initiative.

Caldecott, B., Clark, A., Koskelo, K., Mulholland, E., & Hickey, C. (2021). Stranded assets: Environmental drivers, societal challenges, and supervisory responses. *Annual Review of Environment and Resources, 46*, 417–447.

Carney, M. (2015). Breaking the tragedy of the horizon–climate change and financial stability. *Speech Given at Lloyd's of London, 29*, 220–230.

Carney, M. (2019). Fifty shades of green. *Finance &* Development, pp. 12–15. Washington.

CBI. (2020). Green bonds: The state of the market 2020. *Climate Bonds Initiative*, p. 31.

CBI. (2021). Post-issuance reporting in the green bond market. *Climate Bonds Initiative.*

Chen, Y., & Zhao, Z. J. (2021). The rise of green bonds for sustainable finance: Global standards and issues with the expanding Chinese market. *Current Opinion in Environmental Sustainability, 52*, 54–57.

Cœuré, B. (2018). Monetary policy and climate change, speech by B. Cœuré, member of the executive board of the ECB, at a conference on "scaling up green finance: The role of central banks". *European Central Bank*. https://www.ecb.europa.eu/press/key/date/2018/html/ecb.sp181108.en.html

Coletaon, A., Brucart, M., Gutierrez, P., Le Tennier, F., & Moor, C. (2020). *Sustainable finance—Market practices* (EBA staff paper series [6]).

COP. (2015). Adoption of the Paris Agreement. *UN-Framework Convention on Climate Change*. https://unfccc.int/resource/docs/2015/cop21/eng/l09.pdf

Cunha,F. A. F. d. S., Meira, E., & Orsato, R. J. (2021). Sustainable finance and investment: Review and research agenda. *Business Strategy and the Environment, 30*(8), 3821–3838.

D'Orazio, P. (2022). Mapping the emergence and diffusion of climate-related financial policies: Evidence from a cluster analysis on G20 countries. *International Economics, 169*, 135–147. https://doi.org/10.1016/j.inteco.2021.11.005

D'Orazio, P., & Popoyan, L. (2019). Fostering green investments and tackling climate-related financial risks: Which role for macroprudential policies? *Ecological Economics, 160*, 25–37. https://doi.org/10.1016/j.ecolecon.2019.01.029

D'Orazio, P. (2021). Towards a post-pandemic policy framework to manage climate-related financial risks and resilience. *Climate Policy, 21*(10), 1368–1382. https://doi.org/10.1080/14693062.2021.1975623

D'Orazio, P., & Dirks, M. (2021). Exploring the effects of climate-related financial policies on carbon emissions in G20 countries: A panel quantile regression approach. *Environmental Science and Pollution Research*, 1–25. https://doi.org/10.1007/s11356-021-15655-y

D'Orazio, P., & Popoyan, L. (2022). Realising central banks' climate ambitions through financial stability mandates. *Intereconomics, 57*(2), 103–111.

EBA. (2020a). On management and supervision of ESG risks for credit institutions and investment firms (EBA discussion paper). *European Banking Authority*.

EBA. (2020b, January). *Sustainable finance—Market practices* (n. 6). European Banking Authority.

EC. (2018). *Commission action plan on financing sustainable growth* (Tech. Rep.). European Commission.

Fatemi, A. M., & Fooladi, I. J. (2013). Sustainable finance: A new paradigm. *Global Finance Journal, 24*(2), 101–113.

Fiedler, T., Pitman, A. J., Mackenzie, K., Wood, N., Jakob, C., & Perkins-Kirkpatrick, S. E. (2021). Business risk and the emergence of climate analytics. *Nature Climate Change, 11*(2), 87–94.

Folqué, M., Escrig-Olmedo, E., & Corzo Santamaría, T. (2021). Sustainable development and financial system: Integrating ESG risks through sustainable investment strategies in a climate change context. *Sustainable Development*, *29*(5), 876–890.

FSB. (2022, April 29). *Supervisory and regulatory approaches to climate-related risks—interim report*. Financial Stability Board.

FSCM. (2019). *Mongolian green taxonomy* (Tech. Rep.). Financial Stability Commission of Mongolia.

Gatti, L., Seele, P., & Rademacher, L. (2019). Grey zone in–greenwash out. A review of greenwashing research and implications for the voluntary-mandatory transition of CSR. *International Journal of Corporate Social Responsibility*, *4*(1), 1–15.

GCoA. (2019). *Adapt now: A global call for leadership on climate resilience*. Global Commission on Adaptation—World Resources Institute.

GFC. (2015). *China green bond endorsed project catalogue* (December 2015 ed.) (Tech. Rep.). Green Finance Committee of China Society of Finance and Banking.

GSIA. (2019). *2018 global sustainable investment review*. Global Sustainable Investment Review.

IPCC. (2018). *Global warming of 1.5 degrees: Summary for policymakers* (Tech. Rep.). Intergovernmental Panel on Climate Change. https://report.ipcc.ch/sr15/pdf/sr15spmfinal.pdf

IPCC. (2021). *AR6 climate change 2021: The physical science basis* (Tech. Rep.). Intergovernmental Panel on Climate Change. https://www.ipcc.ch/report/ar6/wg1/

Jebe, R. (2019). The convergence of financial and ESG materiality: Taking sustainability mainstream. *American Business Law Journal, 56*(3), 645–702.

Khan, M. A. (2022). ESG disclosure and firm performance: A bibliometric and meta analysis. *Research in International Business and Finance, 61* , 101668.

Kölbel, J. F., Heeb, F., Paetzold, F., & Busch, T. (2020). Can sustainable investing save the world? reviewing the mechanisms of investor impact. *Organization & Environment*, *33*(4), 554–574.

Krogstrup, S., & Oman, W. (2019). *Macroeconomic and financial policies for climate change mitigation: A review of the literature*. International Monetary Fund.

Kron, W., Löw, P., & Kundzewicz, Z. W. (2019). Changes in risk of extreme weather events in Europe. *Environmental Science & Policy*, *100*, 74–83.

MacAskill, S., Roca, E., Liu, B., Stewart, R. A., & Sahin, O. (2021). Is there a green premium in the green bond market? Systematic literature review revealing premium determinants. *Journal of Cleaner Production, 280*, 124491.

Maiti, M. (2021). Is ESG the succeeding risk factor? *Journal of Sustainable Finance & Investment, 11*(3), 199–213.

McInerney, C., & Bunn, D. W. (2019). Expansion of the investor base for the energy transition. *Energy Policy, 129*, 1240–1244.

Monasterolo, I. (2020). Climate change and the financial system. *Annual Review of Resource Economics, 12*, 299–320.

NGFS. (2017). *Charter of the Central Banks and supervisors network for greening system*. Network for Greening the Financial System.

NGFS. (2019). *A call for action—Climate change as a source of financial risk, first comprehensive report*. Network for Greening the Financial System.

NGFS. (2020). *Survey on monetary policy operations and climate change: Key lessons for further analyses*. Network for Greening the Financial System.

NGFS. (2021). *Biodiversity and financial stability: Building the case for action*.

OECD. (2020a). *Sustainable finance definitions and taxonomies in China in "developing sustainable finance definitions and taxonomies"* (Tech. Rep.). Organization for Economic Cooperation and Development.

OECD. (2020b). *Sustainable finance definitions in France in "developing sustainable finance definitions and taxonomies"* (Tech. Rep.). Organization for Economic Cooperation and Development.

OMFIF. (2020). *Tackling climate change: The role of banking regulation and supervision*. Official Monetary and Financial Institution Forum & Luxembourg for Finance.

OMFIF. (2022). *Forging the path to international standards in sustainable finance*. Official Monetary and Financial Institution Forum & Luxembourg for Finance.

Popescu, I.-S., Hitaj, C., & Benetto, E. (2021). Measuring the sustainability of investment funds: A critical review of methods and frameworks in sustainable finance. *Journal of Cleaner Production, 314*, 128016.

Roy, J., Ghosh, D., Ghosh, A., & Dasgupta, S. (2013). Fiscal instruments: Crucial role in financing low carbon transition in energy systems. *Current Opinion in Environmental Sustainability, 5*(2), 261–269.

Ruiz-Blanco, S., Romero, S., & Fernandez-Feijoo, B. (2022). Green, blue or black, but washing—What company characteristics determine greenwashing? *Environment, Development and Sustainability, 24*(3), 4024–4045.

Sachs, J. D., Woo, W. T., Yoshino, N., & Taghizadeh-Hesary, F. (2019). Importance of green finance for achieving sustainable development goals and energy security. In *Handbook of green finance* (pp. 3–12). Springer.

SBFN. (2022). *Global progress reports 2022*. Sustainable Banking and Finance Network (SBFN). https://www.sbfnetwork.org/publications/global-progress-report-2021/

SBN. (2020, March). *Sustainable Banking Network (SBN) background*. Task Force on Nature-related Financial Disclosures. https://tnfd.global/wp-content/uploads/2022/03/220321-TNFD-framework-beta-v0.1-FINAL.pdf

Steuer, S., & Tröger, T. H. (2022). The role of disclosure in green finance. *Journal of Financial Regulation*, *8*(1), 1–50.

TCFD. (2017). *Final report: Recommendations of the task force on climate-related financial disclosures*. Task Force on Climate-related Financial Disclosures. https://www.fsb-tcfd.org/publications/final-recommendations-report

TCFD. (2018). *TCFD: 2018 status report*. Task Force on Climate-related Financial Disclosures. https://www.fsb-tcfd.org/publications/tcfd-2018-status-report

TCFD. (2021). *TCFD: 2021 status report*. Task Force on Climate-related Financial Disclosures. https://assets.bbhub.io/company/sites/60/2021/07/2021-TCFD-StatusReport.pdf

TEG. (2020). *Taxonomy: Final report of the technical expert group on sustainable finance, March 2020* (Tech. Rep.). EU Technical Expert Group on Sustainable Finance.

TNFD. (2022, March). *The TNFD nature-related risk & opportunity management and disclosure framework (beta v01. release)*. Task Force on Nature-related Financial Disclosures. https://tnfd.global/wp-content/uploads/2022/03/220321-TNFD-framework-beta-v0.1-FINAL.pdf

Uyar, A., Karaman, A. S., & Kilic, M. (2020). Is corporate social responsibility reporting a tool of signaling or greenwashing? evidence from the worldwide logistics sector. *Journal of Cleaner Production, 253*, 119997.

Yu, E. P.-y., Van Luu, B., & Chen, C. H. (2020). Greenwashing in environmental, social and governance disclosures. *Research in International Business and Finance*, *52*, 101192.

The Role of Bank Regulators in the Promotion of Green and Climate Finance

Paola D'Orazio

1 Introduction

Climate change is one of humanity's most significant challenges in the twenty-first century. An increase in the average global temperature of more than 2 °C will imply a rise in the frequency of weather-related catastrophes as well as the slow-moving but potentially destructive processes of ocean acidification and sea-level rise and would threaten the very existence of our species because of damage to the ecosystem. The UN Intergovernmental Panel on Climate Change emphasized that an increase in global carbon emissions brought on by human activity will have "severe, pervasive, and irreversible implications for people and ecosystems" (IPCC, 2014, 2018). Mitigation and adaptation measures should thus be at the forefront of the global policy agenda so emissions could peak to achieve Paris Agreement goals (IPCC, 2022).

P. D'Orazio (✉)
Chair of Macroeconomics, Ruhr-Universität Bochum, Bochum, Germany
e-mail: paola.dorazio@ruhr-uni-bochum.de

C. Gaganis et al. (eds.), *Sustainable Finance and ESG*, Palgrave Macmillan Studies in Banking and Financial Institutions,
https://doi.org/10.1007/978-3-031-24283-0_8

It is generally agreed that governments must take the initiative in solving these problems through climate mitigation and adaptation policies. However, climate change is a systemic risk and endangers the conduct of monetary policy, as well as the stability of the financial system, besides countries' fiscal space (Alexander, 2014; Bolton et al., 2020; D'Orazio & Popoyan, 2019; Krogstrup & Oman, 2019). Along with high economic costs, extreme weather events can indeed cause financial losses for non-financial businesses and exacerbate their financial vulnerability (Carney, 2015). In particular, climate-related financial risks can lead to credit, market, liquidity, and insurance risks due to financial and economic losses, the destruction of production capital, the decline in profitability of exposed firms, and the stranding of assets related to climate-relevant sectors like, for example, mining and fossil fuels (Caldecott et al., 2018, 2021). Therefore, central banks, financial regulators, and supervisors are expected to evaluate how financial institutions consider social and environmental issues and provide guidelines and rules for how their decisions affect the low-carbon transition (D'Orazio & Popoyan, 2019). In particular, to pursue institutional goals related to monetary and financial stability, central banks, financial regulators, and supervisors must thoroughly understand how climate change affects pricing and economic growth over longer periods (Batten et al., 2016; Elderson, 2018; HLEG, 2018).

Over the past years, there has been an increase in interest in including environmental and climate-related issues in the activities of central banks and financial regulators and supervisors (D'Orazio, 2022). The scientific discussion and actions taken by these institutions are presented in this chapter. The review describes the ways and the degree to which central banks are increasingly seen as key players in advancing regulations supporting a low-carbon or "net-zero" agenda. Moreover, it illustrates future policy directions and the new policy tools that central banks and financial regulators need to consider as climate change and its economic consequences are anticipated to generate new risks for financial markets.

The chapter is organized as follows. Section 2 provides context for the analysis. Section 3 reviews climate-related financial policy actions, and the adoption and diffusion of such measures. Section 4 discusses the shortcomings of existing micro- and macro-prudential frameworks. Section 5 examines the so-called green monetary policy debate and existing experiences. Finally, Section 6 provides concluding remarks and highlights future policy directions.

2 Climate Risks: Why Do They Matter for Financial Stability and Monetary Policy?

Following the 2008 financial crisis, central banks were subject to an intensive discourse on their role in safeguarding financial stability and their mandate more broadly; this resulted in intensified efforts to strengthen financial regulation and identify systemic financial risks to mitigate them (Goodhart et al., 2011). More recently, this debate has evolved to address the concerns related to the role of central banks and financial regulators in tackling climate change by addressing climate-related financial risks and contributing to scale up green finance (Campiglio et al., 2018; Cœuré, 2018; OMFIF, 2020; Schnabel, 2020). This attention is motivated by the fact that climate-related financial risks are unique and derive from extreme weather events' strong potential impact and irreversibility. Due to the increasing frequency and severity of storms, floods, heatwaves, or changes in climate patterns, there can be substantial economic costs and financial losses, harming non-financial firms, and increasing their financial fragility. Overall, physical and transition risks will affect the financial sector because of devaluation or write-off of assets from economic agents' balance sheets. Related to this, since the macro-financial system is characterized by networks based on balance sheet interactions, non-financial firms' fragility is transmitted to both the financial and real sectors (Cahen-Fourot et al., 2019).

Additionally, extreme weather events and a sudden transition to a low-carbon economy because of changes in market players' preferences or climate policies can cause asset stranding, i.e., the unanticipated devaluation of carbon-intensive assets for the exposed firms, which has adverse effects on employment, tax revenues, and trade dynamics. For example, in oil-exporting countries, a sudden reduction of their dependency on fossil fuels because of a climate-related event or the need for a sudden transition to a low-carbon economy, imposed, for example, by the adoption of a mitigation policy could imply, among others, a decline in the export revenues and a reduction of domestic consumption, to which will follow severe macroeconomic consequences (Ansari & Holz, 2020).

The literature classifies the risks posed to financial stability by climate change in two main categories: physical and transition risks (Carney, 2015). The former is associated with the economic cost of actual or expected extreme climate events that can cause the erosion and high volatility of physical and financial assets' monetary value, thus increasing overall uncertainty in financial markets. The latter derives from a sudden or disorderly transition, triggered by, among others, unanticipated changes in public policy caused by market participants and concerns about their destabilizing effects on the financial system, such as lower portfolios value, higher non-performing loans in banks' balance sheets, or a decline in returns for insurance companies (Batten et al., 2016).

To address the issues raised by climate-related financial risks to financial stability, central banks and financial regulators worldwide have started to reflect on the tools and instruments at their disposal (de Galhau et al., 2019; Schnabel, 2020; Weidmann, 2019). They range from monetary policy tools, such as allocation decisions when purchasing assets and taking collateral (Matikainen et al., 2017; Schoenmaker, 2021), to "green" micro- and macro-prudential instruments (D'Orazio & Popoyan, 2019).

3 Climate Change and Financial Supervision and Regulation

Central banks and financial regulators' engagement in climate-related action is related, on the one hand, to the goal of scaling up green finance; on the other hand, they are called to assess and tame financial instability.

Scaling up Green Finance

Thanks to the sizeable national pledges made at COP16, climate funds started to flow from developed to developing countries. The estimated climate funds conveyed to developing countries have risen in recent years. However, some scholars have recently pointed out the potential harmful effects, such as market volatility, Dutch disease, rent-seeking, and corruption of climate funds. In contrast, others have emphasized their effectiveness in promoting green growth in the targeted countries (see Carfora & Scandurra, 2019, among others). There is, however, a general claim about the existence of a "green finance gap"; i.e., the lack of adequate financial resources to be directed to green investments, environmental technologies, and eco-innovation needed to achieve the COP21 goal of limiting global warming (Buchner et al., 2017; IEA, 2018; Sachs et al., 2019). Thus, the existing volumes of climate finance fall short of meeting the 2 °C scenario called for by IPCC (2018), and a green structural change is difficult to achieve. Moreover, especially in developing countries, the public sector cannot afford to fill the green investment gap, and the private sector has not shown sufficient interest in engaging in it because of the low rate of return and the associated risks. Thus, several policy actions are required to face the hurdles posed by climate change. Relying on market dynamics alone might be too challenging (Krogstrup & Oman, 2019; Stiglitz et al., 2017); therefore, national jurisdictions have started to consider new financial instruments and policies, such as green bonds (Baker et al., 2018), green investments promoted by state investment banks (D'Orazio & Valente, 2018; Mazzucato, 2015), carbon market instruments (Baranzini et al., 2017), green fiscal policies (Polzin & Sanders, 2020), and green central banking (Campiglio et al., 2018), to scale up green finance and address climate-related financial risks.

Following the Paris Agreement goal (Article 2.1c) of "making finance flows consistent with a pathway towards low greenhouse gas emissions and climate-resilient development" (COP, 2015), central banks and financial regulators can redirect financial flows toward activities that protect natural capital and positively affect the environment, thus enabling their "green and climate finance action." This is particularly relevant for a successful low-carbon just transition because green and climate finance consider environmental protection and the effective use of resources as important criteria for measuring investments' effectiveness and contributes to sustainable development, and promotes economic growth (see, e.g., IPCC, 2021).

Taming Climate-Related Financial Instability

Since the signing of the Paris Agreement, many financial regulators have shown greater engagement in addressing climate-related financial risks and coping with climate uncertainty. Following the seminal contribution of the former Governor of the Bank of England (Carney, 2015), the Financial Stability Board advocated for the Task Force on Climate-related Financial Disclosures creating the High-Level Expert Group (TCFD, 2017). Nevertheless, global financial markets are mostly misaligned with the Paris Agreements' goals and affected by a so-called carbon bias, contributing to carbon lock-in and path dependence and implying potential destabilization threats.

Financial Policy Area	Description	Category		Instrument	Objective	Countries that adopted (authority responsible for promotion/implementation)
Policy Area I	**Green Prudential Regulations:** to promote the development of green macroprudential frameworks	*Capital*	*Quality and level of capital*	CAR with GSF/BPF CCyB Sectoral Leverage Ratios Sectoral Capital Requirements	Mitigate and prevent excessive credit growth and leverage	
			Risk management and supervision	Governance and risk management		**Australia (PRA), China (CB, PRA, GOVT), France (PRA, GOVT), Indonesia (GOVT), Mexico (CB), UK (PRA)**
				Climate-related stress test (macro)	Evaluate effect of economic or financial shocks to the financial system	**Canada (CB), China (PRA), France (PRA), UK (PRA)**
				Green Asset Ratio	Assess exposure of of banks' portfolios to carbon-intensive assets	
				ICAAP	Internal Process of Capital Adequacy Assessment: Include social and environmental risks when assessing their capital needs	**Brazil (CB)**
			Enhanced risk disclosure and market discipline	Climate-related disclosure requirements	Inform about concentration of carbon-intensive assets in the financial sector	**China (PRA), Indonesia (CB), Mexico (BA), Turkey (BA)**
		Liquidity	*Liquidity*	LCR NSFR	Mitigate and prevent market illiquidity and maturity mismatch	
		Large exposures	*Lending limits*	Large exposures limit	Mitigate systemic risk by limiting the concentration of certain exposures	
Policy Area II	**Green Financial Principles:** to create green financial markets					**All G20 countries except Saudi Arabia**
Policy Area III	**Other disclosure requirements:** to promote the public disclosure of climate risks (also for non-financial institutions)					**All G20 countries except Argentina, Mexico, Saudi Arabia, Turkey**
Policy Area IV	**Green bonds taxonomy and issuing:** to promote the development of green financial securities					**Indonesia (Other)**
Policy Area V	**Green Credit Allocation Policies:** to directly promote green credit measures and investments					**Japan (GOVT), South Korea (GOVT)**

Legend:
BA: BANKING ASSOCIATION
PRA: PRUDENTIAL REGULATORY AUTHORITY
CB: CENTRAL BANK
GOVT: GOVERNMENT
OTHER: ISTITUTION OTHER THAN THE PREVIOUS CATEGORIES

Fig. 1 Overview of the five policy areas considered in the analysis (*Source* D'Orazio and Thole [2022])

Adoption of Climate-Related Financial Policies

Existing evidence shows that the adoption of climate-related financial policies (CRFPs)[1] has grown steadily over the past twenty years, reflecting an increased engagement of countries globally, as reported in Fig. 1.

Figure 2 shows that by the end of 2020, green finance guidelines, green bonds, prudential policies, and credit allocation measures—for which only ten implemented measures have been identified—were the most often reported policies. The methodology developed in D'Orazio and Thole (2022) reveals, however, that a high climate-related financial policy index (CRFPI) characterizes very few countries: Australia, Brazil, China, France, Indonesia, the Netherlands, and South Korea.[2] Other countries at the bottom of the list with rankings below average, like Saudi Arabia, Argentina, Turkey, Italy, Canada, Russia, and the United States, need to engage in climate-related financial policymaking if they are to meet the goals of the Paris Agreement. 38% of the global CRFPs adopted globally have been promoted by central banks and financial supervisors. By computing the index on the subset of policies promoted exclusively

[1] Following existing literature (D'Orazio, 2021, 2022; D'Orazio & Thole, 2022; Krogstrup & Oman, 2019), the climate-related financial policies considered in this study include five policy areas as described in Fig. 1.

[2] The CRFPI is a composite index for assessing, quantifying, and comparing international engagement in climate-related financial policymaking. For a comprehensive description of the methodology and results, see D'Orazio and Thole (2022).

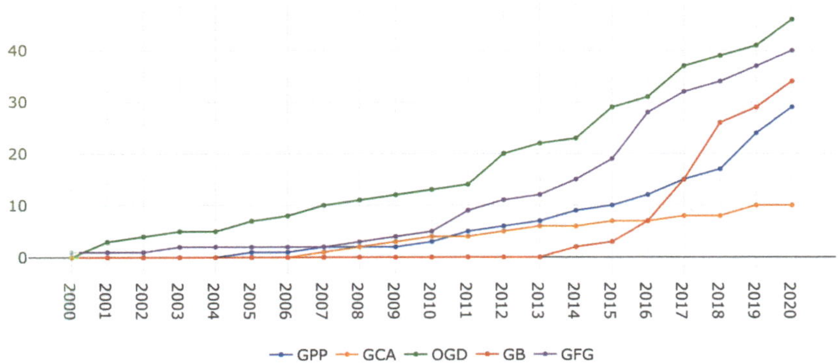

Fig. 2 Total number of policies adopted yearly at the global level; time span: 2000–2020 (*Notes* GPP: Green Prudential Policy; OGD: Other Green Disclosure Req.; GFG: Green Financial Guidelines; GB: Green Bonds; GCA: Green Capital Allocation. *Source* D'Orazio and Thole [2022])

by central banks and financial regulators, it emerges that (i) the most committed country to green financial policymaking is Indonesia, followed by Brazil and then China and the UK, and (ii) the majority of the countries have a policy performance that is below the index average.

Drivers of Diffusion of Climate-Related Financial Policies
Regarding the factors that influence CRFPs' adoption, D'Orazio (2022) has shown that a number of variables influence the diffusion process and that three waves of policy adoption can be observed. Higher emissions, i.e., an economy's carbon intensity and exposure to climate change, result in a higher adoption rate; however, the financial policy response to these factors varies across countries. In particular, it is found that only countries classified in the second adoption wave (starting in 2008 and including China and other Asia-Pacific countries) show some predisposition to policymaking tackling climate risks. Countries in the first wave—which started in 2000, with the leading example of European countries—adopted mainly financial principles and disclosure requirements for non-financial institutions. The third wave, instead, is more recent (started in 2012–2013) and characterizes Central and East Asian economies as more oriented to "softer" policy types. Political characteristics, particularly the presence of an autocratic regime defined by the

polity index, are relevant for the Chinese case, calling "authoritarian environmentalism" to mind. Finally, a bandwagon, or geographical learning effect, plays a relevant role in all clusters. Climate strategies (i.e., formal and legally binding strategy or a political and non-binding strategy) and fiscal instruments—such as feed-in-tariffs and carbon pricing schemes—are relevant only for first-wave adopters in high-income countries, such as France and the UK. In the case of China, political traits, particularly the existence of an autocratic regime as denoted by the polity index, are significant and bring to mind "authoritarian environmentalism." Finally, a bandwagon effect, also known as the geographical learning effect, is significant.

Waves of Adoption of Climate-Related Financial Policies

The empirical analysis conducted in D'Orazio (2022) shows the existence of six defined clusters, as reported in Fig. 3. Their differences relate to the timing and the different shares of adopted financial policy types. Two clusters are identified in the first wave of adoption, namely Clusters I and II. Two clusters can be classified in the second wave: Cluster III and V. Finally, a group of latecomers is observed in Cluster IV and VI. An overview of the three waves and the clusters belonging to each wave is provided in Fig. 4.

The findings emphasize that (i) first-wave adopters focus mainly on policy types III and IV; (ii) second-wave adopters focus on a richer set

		CLUSTER I		CLUSTER II		CLUSTER III		CLUSTER IV		CLUSTER V		CLUSTER VI	
		polity	# policies adopted	polity	# policies adopted	polity	# policies adopted	polity	# policies adopted	polity	# policies adopted	polity	# policies adopted
		France	9	Brazil	9	Australia	9	India	4	China	8	Russian Federation	5
		UK	8	Indonesia	5	Japan	4	Indonesia	1	Indonesia	1	Turkey	2
		Italy	6	Mexico	3	Korea, rep.	4			South Africa	1		
		Germany	4	South Africa	2	USA	3						
		Canada	2	Argentina	1	Canada	1						
		Saudi Arabia	1										
	Policy adoption: total		30		20		21		5		10		7
	Policy adoption: share		32.25%		21.5%		21.58%		5.37%		10.7%		7.52%
Adoption by type: share	Policy type I		1.07%		2.15%		1.07%		0		6.54%		1.07%
	Policy type II		1.07%		5.37%		2.15%		1.07%		0		0
	Policy type III		12.9%		9.6%		12.9%		0		2.15%		3.22%
	Policy type IV		17.2%		4.3%		6.54%		4.3%		2.15%		3.22%

Fig. 3 Climate-related financial policymaking in G20 countries: cluster structure, number of policies, and share of policy adoption by type (*Source* D'Orazio [2022])

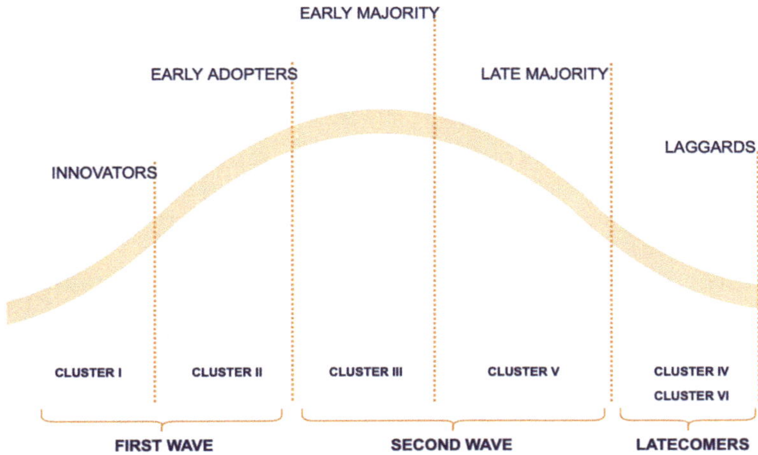

Fig. 4 Overview of the different waves of adoption (*Source* D'Orazio [2022])

of policies and display the highest (aggregate) share of policy type I; (iii) latecomers have a stronger preference for policy type IV.

An important "bandwagon," or geographical learning effect, is identified in the dynamics observed in the sample. The empirical analysis shows that countries are more likely to engage in climate-related financial policymaking when the share of countries undertaking reforms on the same continent or region is high. In particular, we detect three different phases in which the bandwagon effect has been at work. The first phase of the first wave started in 2000, with the leading example of European countries (Cluster I) adopting principally policy types III and IV. A second phase started in 2005 in Indonesia and "propagated" to South and East Asia, followed in 2008 by Latina American countries, under Brazil's leading example. A second wave is led by China, which adopted the first climate-related financial policy in 2007; other Asia-Pacific countries—such as Japan, South Korea, and Australia—succeeded in the following years. Finally, a third wave is more recent, starting in 2012–2013, and characterizes Central and East Asian economies.

4 Challenges and Shortcomings of Existing Prudential Frameworks

Assessing the Current Micro-Prudential Framework

Analyzing the inclusion of ESG perspectives in the Pillars of the Basel III framework, evidence shows that they do not handle crucial aspects of climate risks, namely the cross-sectoral, global, and systemic dimensions. The motivation is twofold.

First, existing tools consider risks that manifest over a shorter time frame than climate threats. As a result, climate risks are not fully captured and are only indirectly reflected at best.

Second, by focusing on historical losses, the methods employed to evaluate the risks fail to represent the "fundamentally uncertain nature" of climate hazards. In terms of Pillar 1, existing regulations do not require banks to examine the impact of climate-related risks on their exposures (see D'Orazio & Thole, 2022, for a recent review). Consequently, they encourage carbon bias and short-termism in financial markets, making capital mobilization more difficult for green investment projects. Moreover, current approaches do not allow for appropriate calibration of climate-related hazards (BoE-PRA, 2021; Coelho & Restoy, 2022); therefore, a dedicated prudential treatment of such risks would be appropriate (Chenet et al., 2021; EBA, 2022b).

Some proposals have highlighted the possibility of limiting the carbon bias and possibly increasing the share of low-carbon investments by using a "Brown Penalizing Factor" (BPF) to calculate banks' Capital Adequacy Requirements (2DII, 2018). Others have recommended a Green Supporting Factor (GSF), which proposes to lower the capital requirement for green assets (Dombrovskis, 2017). However, policymakers and academic researchers have heavily criticized these proposals as they could lead to severe market distortions and potential financial instabilities, thus contradicting the original aim of the measure (D'Orazio & Popoyan, 2019).

Moreover, adjusting risk weights set by the regulators to be used in the Standard Approach (SA) or in the Internal Risk-Based (IRB) approach to estimate the risk-weighted assets (RWA) and then revise the capital requirements accordingly is still seen as a costly task for supervisors and regulators in terms of research efforts, resources to build new expertise and regulatory adaptation plans. In the European Union, a recent proposal for reform of the Capital Requirement Regulation (CRR) by the European Banking Authority (EBA, 2020, 2022a) stated that it would assess "whether a dedicated prudential treatment of exposures related to assets or activities associated substantially with environmental and/or social objectives would be justified." A report on this matter will be delivered by June 2025, and after that, the EU Commission could decide to submit a legislative proposal to the European Parliament and the Council. However, the timeline for such reform is delayed in time, thus making this option very unlikely to be implemented in the short term.

Because of the costly and potentially long (i.e., time-consuming) reform of Pillar 1 to include the climate and environmental risks in the capital requirements (BoE-PRA, 2021; ECB, 2020b), it is often argued that supervisors and regulators could rely on Pillar 2 to implement the required changes (see, e.g., Coelho & Restoy, 2022). The argument is that climate-related scenario analysis and stress tests can be used to assess the impact of climate risks on banks' balance sheets. This, in turn, will increase their awareness about exposure to climate risks and possible deficiencies in risk management practices through the implementation of an internal capital adequacy assessment process (ICAAP). The standard Pillar 2 is considered "more flexible" compared to Pillar 1 as it allows supervisors to require financial institutions to change the management approach to risks and create additional loss-absorption capacity (i.e., a capital add-on) when deficiencies (in the management of risks defined under Pillar 1) are found (Coelho & Restoy, 2022). In the same vein, Frank Elderson proposed that banks be required to establish Paris-compatible transition plans as a legal requirement[3] (Elderson, 2021).

Climate-related stress testing is one of the most important instruments under Pillar 2 and can be utilized for both micro and macro-prudential purposes. They aim to identify how resilient the financial system is to

[3] They "[...] should highlight at any given point in time, from now until 2050, the bank's alignment and potential divergences with the relevant policy objectives through which the EU implements the Paris Agreement. [...] should be part of a bank's strategy-setting and be closely linked to its business model and business plan. It should contain concrete intermediate milestones from now until 2050 and the associated key and performance indicators so that the bank's management and the competent authorities can understand the risks arising from a possible misalignment with the transition path. If banks fail to meet these milestones, competent authorities – including prudential supervisors – will have to take appropriate measures to ensure that this failure does not result in financial risks."

adverse climate shocks by looking at the impact of hypothetical climate-related shock scenarios on individual financial institutions (the former) and the financial system (the latter). They also provide policymakers with essential information on the financial system's exposure to climate-related risks, and their findings might be used to calibrate and evaluate climate-related macro-prudential measures. However, despite their importance, only very few countries, namely Canada, China, France, and the UK, have actively considered climate-related financial risks through stress tests.

Regarding Pillar 3, the current debate is focused on enhancing disclosure standards and making them mandatory. The focus is on ESG risk disclosures, which are considered critical to foster market discipline (i.e., Pillar 3's core). Risk disclosures are relevant to inform market participants about concentrations of carbon-intensive assets at the portfolio level, thus allowing stakeholders to assess banks' ESG-related risks and sustainable financing strategies. However, if they are not considered alongside other prudential policies and made mandatory at the international level, they may not be enough to meet the challenges of climate finance alignment (Ameli et al., 2019). The current debate and action on financial disclosure are quite advanced compared to other areas of regulation and benefits from the research and recommendations of the TCFD (2017). Despite the debate being quite advanced on disclosure requirements, they are mandatory for financial institutions only in China, Indonesia, Mexico, and Turkey, among G20 countries (see D'Orazio & Thole, 2022; D'Orazio, 2021, for a recent review of the adoption of climate-related financial policies). The EBA has recently issued a report on Pillar 3 disclosures on ESG risks that propose new standards or modifications to existing measures (Coletaon et al., 2020). Among the new measures, a Green Asset Ratio (GAR) on Taxonomy-aligned activities is proposed. Since it is also used in other policy initiatives to understand institutions' exposures to environmentally sustainable activities, it is considered particularly useful in this framework. However, it might be argued that the GAR might not be an adequate tool to measure the alignment of banks' portfolios to low-carbon transition and falls better under Pillar 2.

Assessing the Current Macro-Prudential Framework

The current financial policy framework is insufficient to assess the system's vulnerability to climate-related financial risks or redirect financial flows to sustainable investments (D'Orazio, 2021; D'Orazio &

Thole, 2022). Additionally, pandemic-related macro-prudential financial regulations may have exacerbated existing climate-related vulnerabilities (D'Orazio, 2021). The argument is that failing to consider climate change or green finance could encourage more lending to carbon-intensive industries, reinforcing the so-called (already high) "carbon bias." In this environment, countries' overall exposure to climate-related financial risks might further increase, potentially jeopardizing the transition to a low-carbon economy. Given the systemic nature of climate risks, macro-prudential measures should not be overlooked and should be given special attention. Micro-prudential tools, as outlined in section "Assessing the Current Micro-Prudential Framework", are typically focused on direct exposures and may not be sufficient to address the systemic dimension of climate hazards. As a result, macro-prudential instruments must be used in conjunction with micro-prudential instruments. Sectoral exposures and leverage ratios, among other techniques, should be carefully studied for implementation. Existing measures, such as (systemic) capital requirements, do not address climate and environmental risks. The reasons are consistent with the examination of existing micro-prudential capital tools described in section "Assessing the Current Micro-Prudential Framework" and usually point to a lack of sufficient evidence to trigger a risk factor adjustment that penalizes carbon-intensive assets while "promoting" low-carbon assets.

The discussion over climate-related macro-prudential tools usually focuses on the impact of climate change on credit risks and the revision of capital instruments, and the risk factors that must be addressed. Liquidity risks are frequently disregarded, but the potential negative impact of these risks should not be overlooked, and policymakers should explore them further (D'Orazio, 2021; D'Orazio et al., 2022). If we consider the occurrence of a severe weather event, what will be the non-financial agents' (i.e., households and firms) reactions? Following such an occurrence, households may prefer to withdraw funds from their bank accounts (causing a bank run), and companies may decide to rely less on external financial resources (such as bank loans), which may prove extremely expensive. As a result of this climate-induced behavioral response, banks' liabilities are affected because their access to stable funding (deposits) may be reduced. As banks are embedded in a network and interact in the interbank market to replenish their funding sources, interbank exchanges may create two funding-lending cycles. For example, banks may decide to fund short-term their long-term assets, stimulating green investments while damaging their balance sheets' maturity structure. Alternatively, they could employ long-term funds to provide short-term loans, favoring bank stability at the expense of long-term credit provision to favor the low-carbon transition. Furthermore, asset stranding may cause a revaluation of those assets, generating substantial funding and market liquidity shortages for the financial institutions holding the assets and other institutions connected to them through the banking network. Finally, the transition to a low-carbon economy may lead to the liquidation of some banks' balance sheets, signaling a shift in asset prices (i.e., a price fall) and margin calls, resulting in liquidity issues.

5 Climate Change and the Conduct of Monetary Policy

The consequences of climate-related supply price shocks, market volatility, and economic growth—all linked to inflation through credit spreads, saving rates, and real interest rates (Mukherjee & Ouattara, 2021)—make it risky to implement monetary policy without taking action on climate change from central banks and financial regulators (see, e.g., Cœuré, 2018; Schnabel, 2020). In addition, the physical and transition risks are frequently mentioned when discussing the mechanisms of climate change risk transmission to the financial sector (Batten et al., 2016; Carney, 2015). On the one hand, physical risks can reduce the value of financial organizations' collateral assets and insurance liabilities when they arise, thus directly undermining financial stability (FSB, 2020). On the other hand, transition risks may impact financial markets as a result of, among other things, unanticipated changes in (climate) policy, technological advancements, or shifts in public opinion (BCBS, 2021b).

Although some countries are very active in implementing climate-related financial policy (D'Orazio, 2022), monetary authorities frequently do not pursue green monetary policy because they are concerned about breaking the principles of market neutrality and find it challenging to expand their authority beyond existing mandates (de Galhau et al., 2019; van't Klooster & Fontan, 2020; Weidmann, 2019). The argument is that any intervention that favors green investments results in inefficient capital allocation, including the preference for purchasing green assets and the design of new instruments by the monetary and regulatory authorities. Since many central banks, especially in developed and high-income countries, possess significant institutional and operational independence, addressing long-term sustainability issues is regarded with hesitance (Boneva et al., 2022; NGFS, 2020). Moreover, the lack of adequate data, standardized taxonomies, and disclosure regulations prevent an accurate assessment of the impact of climate risks on central banks' balance sheets (ECB, 2020a, 2020b; FSB, 2020; OMFIF, 2022; Schoenmaker, 2021; Steuer & Tröger, 2022).

The market neutrality ideal is often brought to the discussion to stress that central banks must avoid distorting financial markets to fulfill the policy objectives of (primarily) price and financial stability (van't Klooster & Fontan, 2020). It is argued that their interference in advocating low-carbon activities without a specific mandate could jeopardize their independence and credibility, leading to an institutional impasse. Because doing so would entail politicizing monetary policy, central banks embrace the principle of market neutrality and are reluctant to engage in green monetary policy activities. As discussed in D'Orazio and Popoyan (2022), central banks' green monetary "activism" is often considered the second-best intervention compared to other policy actions, such as taxation of carbon emissions and cap-and-trade policies. It is often argued that incorporating sustainability objectives into the monetary policy's operation may overstretch the mandate, thus creating conflicts between the objectives and endangering institutional independence. However, it is often emphasized that such actions are a massive departure from the non-distributional and market-neutral principle of monetary policymaking (Cochrane, 2020; Olovsson, 2018; Weidmann, 2020).

So far, only a few countries have decided on considerable climate action on the monetary policy side. Since 2018, the People's Bank of China (PBoC) has updated and expanded the list of collaterals it accepts for medium-term loans. According to the PBoC, acceptable collateral includes green bonds, loans, and asset-backed securities from commercial banks with a double-A rating or higher. Since 2012, the same region has benefited from lines of credit sponsored by the "Loan Support Programme," which aims to offer preferential liquidity at reduced interest rates to financial institutions lending to socially and environmentally responsible enterprises. In order to provide priority to environmentally friendly companies, other countries, including Brazil, India, and Indonesia, have implemented credit allocation rules like green lending quotas and concessional loans. The ECB demonstrated a stronger level of engagement in this area by the end of 2020, recognizing climate change as "mission-critical" and underlining the consequences of climate change for the main goal of monetary policy (ECB, 2020c, 2022; Elderson, 2022)

6 Future Research and Policy Directions

Banks have started recognizing sustainability risks due to pressure from customers, investors, and financial regulators. They have also started to support the shift to a more sustainable economy by incorporating sustainability factors into their risk management models and governance frameworks. However, the evidence and discussion reported in this chapter show that global financial markets are mostly misaligned with the Paris Agreements' goals and affected by a so-called carbon bias, contributing to carbon lock-in and path dependence and implying potential destabilization threats. Moreover, banks continue to underreport their exposure to climate risks, including whether they significantly influence their risk profile and how the transition and physical risks affect their business model. In particular, current achievements at the global and national levels fall short of what is required to manage financial risks associated with climate change and promote green finance; the public sector as a whole must increase its commitment to better play its crucial role in facilitating a long-term just and low-carbon transition.

The review of existing evidence presented in this chapter highlights that no climate-related macro-prudential measure concerning capital requirements, leverage ratios, or systemically important banks or liquidity requirements have been adopted in G20 countries. However, significant action has been detected concerning climate-related stress testing. Other policies, such as climate-related disclosure requirements of the climate-related financial risks associated with climate change, are also relevant to developing a credible green financial system and avoiding so-called greenwashing. The Chinese macro-prudential authority, the Indonesian central bank, Turkey, and Mexico's banking associations have promoted banks' disclosure requirements. Instead, disclosure requirements for non-financial institutions, pension funds, insurance companies, and green finance principles, and guidelines have been widely adopted over the past 20 years in most G20 countries. At the euro area level, most countries have developed green market-shaping policies and adopted disclosure requirements for non-financial firms, insurance companies, or institutional investors.

It has also been discussed the extent to which current macro- and micro-prudential regimes are not in line with the objectives of the Paris Agreement because important cross-sectoral, global, and systemic issues are not addressed. As a result, the potential function of green finance as a catalyst is constrained, and climate risks are only partially reflected.

Because of the endogeneity of risk and its related uncertainty, the traditional approach to financial risk, which involves assessing expected values and risk using historical market prices and estimating the probability of defaults, is insufficient for addressing climate risks (Bolton et al., 2020). Indeed, assessing the bank portfolio's exposure to such risks and appropriately assessing the credit risk represented by the assets held (or held in the future) on its balance sheet is the main issue posed by climate hazards to financial stability. This evaluation necessitates the creation of two components: (i) new (forward-looking) risk assessment procedures that consider a longer time horizon than traditional macroeconomic exercises and (ii) methods that allow credit quality to be reflected alongside climate risk exposure.

Regarding the former, new methods may imply adding climate-related hazards and possible policy and technical shocks and shifts in market and customer attitudes toward banks' normal risk scenarios. Among others, climate value-at-risk (Battiston & Monasterolo, 2019; Dietz et al., 2016), scenario analyses and stress tests are examples of forward-looking methods that are used to project risks in the future, as they can assist in quantifying tail risks and clarify the uncertainties inherent to climate-related risks (BCBS, 2021a). However, stress testing and scenario analyses should be mandatory to encourage banks' alignment with the Paris Agreement's targets, as they are critical instruments for assessing direct exposures to climate risks. Regarding the latter, new methodologies are needed to see if any economic sectors or activities (e.g., under the EU taxonomy) have (combined) reduced financial and credit risks. The reasoning is that a risk weight linked with the taxonomy may not be sufficient and may cause significant distortions. Indeed, receiving a green label according to the taxonomy does not imply that the asset is risk-free. We believe that sector and economic activity evaluations might be conducted to acquire evidence that including ESG factors reduces financial and credit risk. This analysis would allow for a more thorough identification of "safe" assets that may qualify for lower capital requirements (in the "spirit" of the GSF).

Capital requirements may be implemented to incentivize financial institutions to invest in green, carbon-neutral assets on the macro-prudential side. Without globally standardized taxonomies and disclosure procedures, changing capital requirements to reward green assets or penalize polluting assets would present significant challenges for policymakers and harm the financial system's resilience (D'Orazio & Popoyan, 2019). Therefore, more work is needed to implement more "audacious" policies to encourage green investments and funding. The existing literature shows that the substantial reforms to the risk weighting approach and capital requirements estimation (capital adequacy ratios) are hampered by practical and political barriers, making them difficult to implement in the short term. Evidence suggests that banks and regulators have changed exposure risk weights in the past to accommodate new data or pursue political goals (see, e.g., EBA, 2016).

Among Pillar 1 measures, sectoral capital requirements can be considered as an alternative to standard capital requirements by considering the results of systemic stress tests and scenarios analyses or the outcomes of other analyses aimed at measuring the carbon intensity of loans by sector of economic activity (Faiella & Lavecchia, 2020) or by geographical location. They could also imply that low-carbon vs. carbon-intensive sectors' risk weights or technology within sectors should be differentiated. A bank's exposure to carbon-intensive sectors could be limited by increased risk weights or higher capital buffers. Limiting over-leverage in carbon-intensive sectors strengthens the system and indirectly reorients loans to non-polluting sectors. However, they assume that bank capital costs will rise to penalize polluting companies, which could generate market distortions in the short term. Furthermore, data suggests that evaluating exposures at a sectoral level may underestimate the total CO2 emissions across a company's whole value chain, making implementation more difficult (FSB, 2020). As a result, further research is required before policymakers can effectively employ this tool. Therefore, additional research in the field of input-output analysis might be useful in this respect.

Compared to the instruments listed above, a sectoral leverage ratio may be a more transparent and simpler. It would be based on determining the bank's capital exposure to assets associated with carbon-intensive sectors, which should be limited to a particular percentage of total assets, with the exact percentage defined by the regulator (D'Orazio & Popoyan, 2019). Like sectoral capital restrictions, this strategy could be especially successful in controlling financial market instabilities because it indirectly inhibits over-leveraging in polluting industries and reorients financial flows toward green ones. However, the implementation feasibility of these instruments is related to granular loan and climate data availability.

Another capital measure to consider is a climate-related countercyclical capital buffer (CR-CCyB), which can be used to promote financial stability in the transition from a high-carbon to a low-carbon economy by assisting banks in avoiding the build-up phase of the carbon-intensive credit cycle (for more information, see D'Orazio & Popoyan, 2019). If correctly calibrated, this measure can mitigate the instability resulting from a disorderly transition. During periods of excessive carbon-intensive credit expansion, it will be activated to strengthen financial institutions' resilience throughout the carbon-intensive credit cycle's upswing. However, because it relates to the still-debated taxonomy characterization of carbon-intensive/polluting activities, the carbon-intensive credit cycle definition is a major challenge for this measure.

Our analysis suggests that liquidity risks should be constantly monitored—and addressed—alongside credit risks. Because of the issues highlighted in the previous sections, existing liquidity measures like the Liquidity Coverage Ratio (LCR) and the Net Stable Funding Ratio (NSFR) should be adjusted to account for potential maturity mismatches from low-carbon long-term investments and to steer low-carbon long-term investments.

Further work should also be put toward improving mandatory and globally standardized disclosure requirements. Although some utilize their definitions, financial institutions use a range of international frameworks and standards to identify environmental, social, and governance (ESG) elements. This demonstrates the lack of consistency in ESG considerations and poses issues with the regulatory mandate: If financial institutions employ different definitions of ESG criteria, risk management, and disclosure results may vary. However, if they are not assessed in conjunction with other prudential laws and made obligatory and uniform at the international level, disclosure requirements could not be adequate to address the issues with climate financing alignment.

References

2DII. (2018). *The green supporting factor. Quantifying the impact on European banks and green finance* (Tech. Rep.). 2 Degrees Investing Initiative.

Alexander, K. (2014). Stability and sustainability in banking reform: Are environmental risks missing in Basel III. *CISL & UNEPFI: Cambridge and Geneva*.

Ameli, N., Drummond, P., Bisaro, A., Grubb, M., & Chenet, H. (2019). Climate finance and disclosure for institutional investors: Why transparency is not enough. *Climatic Change , 160*, 1–25.

Ansari, D., & Holz, F. (2020). Between stranded assets and green transformation: Fossil-fuel-producing developing countries towards 2055. *World Development, 130*, 104947.

Baker, M., Bergstresser, D., Serafeim, G., & Wurgler, J. (2018). *Financing the response to climate change: The pricing and ownership of us green bonds* (Tech. Rep.). National Bureau of Economic Research.

Baranzini, A., Van den Bergh, J. C., Carattini, S., Howarth, R. B., Padilla, E., & Roca, J. (2017). Carbon pricing in climate policy: Seven reasons, complementary instruments, and political economy considerations. *Wiley Interdisciplinary Reviews: Climate Change, 8*(4), e462.

Batten, S., Sowerbutts, R., & Tanaka, M. (2016). *Let's talk about the weather: The impact of climate change on central banks* (Tech. Rep.). Bank of England.

Battiston, S., & Monasterolo, I. (2019). How could the ECB's monetary policy support the sustainable finance transition? *FINEXUS: Center for Financial Networks and Sustainability* .

BCBS. (2021a). *Climate-related financial risks-measurement methodologies* (Tech. Rep.). BCBS—Basel Committee on Banking Supervision—Bank for International Settlements.

BCBS. (2021b). *Climate-related risk drivers and their transmission channels* (Tech. Rep.). BCBS—Basel Committee on Banking Supervision—Bank for International Settlements.

BoE-PRA. (2021, October 28). Climate-related financial risk management and the role of capital requirements, climate change adaptation report 2021. *Bank of England—Prudential Regulation Authority*.

Bolton, P., Despŕes, M., da Silva, L.A.P., Samama, F., & Svartzman, R. (2020). *The green swan*. BIS Books .

Boneva, L., Ferrucci, G., & Mongelli, F. P. (2022). Climate change and central banks: What role for monetary policy? *Climate Policy, 22*, 1–18.

Buchner, B., Oliver, P., Wang, X., Carswell, C., Meattle, C., & Mazza, F. (2017). *Global landscape of climate finance 2017* (Tech. Rep.). Climate Policy Initiative.

Cahen-Fourot, L., Campiglio, E., Dawkins, E., Godin, A., & Kemp-Benedict, E. (2019). *Capital stranding cascades: The impact of decarbonisation on productive asset utilisation* (Ecological Economic Papers; No. 18).

Caldecott, B., Clark, A., Koskelo, K., Mulholland, E., & Hickey, C. (2021). Stranded assets: Environmental drivers, societal challenges, and supervisory responses. *Annual Review of Environment and Resources, 46*, 417–447.

Caldecott,B., Dericks, G., Bouveret, G., Schumacher, K., Pfeiffer, A., Tulloch, D., Kruitwagen, L., & McCarten, M. (2018). *Asset-level data and the energy transition: Findings from ET risk work package 2*, p. 19. University of Oxford, Sustainable Finance Programme.

Campiglio, E., Dafermos, Y., Monnin, P., Ryan-Collins, J., Schotten, G., & Tanaka, M. (2018). Climate change challenges for central banks and financial regulators. *Nature Climate Change, 8*(6), 462.

Carfora, A., & Scandurra, G. (2019). The impact of climate funds on economic growth and their role in substituting fossil energy sources. *Energy Policy, 129*, 182–192.

Carney, M. (2015). Breaking the tragedy of the horizon–climate change and financial stability. *Speech given at Lloyd's of London, 29*, 220–230.

Chenet, H., Ryan-Collins, J., & van Lerven, F. (2021). Finance, climate-change and radical uncertainty: Towards a precautionary approach to financial policy. *Ecological Economics, 183* , 106957. https://www.sciencedirect.com/science/article/pii/S092180092100015X. https://doi.org/10.1016/j.ecolecon.2021.106957

Cochrane, J. (2020). *Central banks and climate: A case of mission creep*. Hoover Institution.

Coelho, R., & Restoy, F. (2022, February). *The regulatory response to climate risks: Some challenges, FSI brief, no. 16*. Financial Stability Institute—Bank for International Settlements.

Cœuré, B. (2018). *Monetary policy and climate change, speech by B. Coeur'e, member of the executive board of the ECB, at a conference on "scaling up green finance: The role of central banks"*. European Central Bank. https://www.ecb.europa.eu/press/key/date/2018/html/ecb.sp181108.en.html

Coletaon, A., Brucart, M., Gutierrez, P., Le Tennier, F., & Moor, C. (2020). *Sustainable finance—Market practices* (EBA staff paper series [6]).

COP. (2015). *Adoption of the Paris agreement*. UN-Framework Convention on Climate Change. https://unfccc.int/resource/docs/2015/cop21/eng/l09.pdf

de Galhau, V., et al. (2019). Climate change: Central banks are taking action. *FSR*, p. 7.

Dietz, S., Bowen, A., Dixon, C., & Gradwell, P. (2016). "climate value at risk" of global financial assets. *Nature Climate Change, 6*(7), 676.

Dombrovskis, V. (2017). *Greening finance for sustainable business*. Speech by Vice-President for the Euro and Social Dialogue, Financial Stability and Financial Services Valdis Dombrovskis (SPEECH/17/5235).

D'Orazio, P. (2022). Mapping the emergence and diffusion of climate-related financial policies: Evidence from a cluster analysis on G20 countries. *International Economics, 169*, 135–147. https://doi.org/10.1016/j.inteco.2021.11.005

D'Orazio, P., & Popoyan, L. (2019). Fostering green investments and tackling climate-related financial risks: Which role for macroprudential policies? *Ecological Economics, 160*, 25–37. https://doi.org/10.1016/j.ecolecon.2019.01.029

D'Orazio, P., & Popoyan, L. (2022). Realising central banks' climate ambitions through financial stability mandates. *Intereconomics, 57*(2), 103–111.

D'Orazio, P., Reale, J., & Pham, A. D. (2022). *Climate-induced liquidity crises: Interbank exposures and macroprudential implications* (Ruhr Universi¨at Bochum Working Paper).

D'Orazio, P., & Thole, S. (2022). Climate-related financial policy index: A composite index to compare the engagement in green financial policymaking at the global level. *Ecological Indicators, 141*, 109065.

D'Orazio, P., & Valente, M. (2018). The role of finance in environmental innovation diffusion: An evolutionary modeling approach. *Journal of Economic Behavior & Organization* . http://www.sciencedirect.com/science/article/pii/S0167268118303457. https://doi.org/10.1016/j.jebo.2018.12.015

D'Orazio, P. (2021). Towards a post-pandemic policy framework to manage climate-related financial risks and resilience. *Climate Policy, 21*(10), 1368–1382. https://doi.org/10.1080/14693062.2021.1975623

EBA. (2016). *European banking authority report on SMES and SME supporting factor*. eba/op 2016/04. European Banking Authority.

EBA. (2020). *Action plan on sustainable finance*. European Banking Authority.

EBA. (2022a). *Implementing technical standards (its) on pillar 3 disclosures on environmental, social and governance (ESG) risks*. European Banking Authority.

EBA. (2022b). *The role of environmental risks in the prudential framework* (Discussion paper EBA/DP/2022b/02, 2 May 2022b, European Banking Authority).

ECB. (2020a). *ECB report on institutions' climate-related and environmental risk disclosures* (Tech. Rep.). European Central Bank—Banking Supervision.

ECB. (2020b). *Guide on climate-related and environmental risks. supervisory expectations relating to risk management and disclosure*. European Central Bank.

ECB. (2020c, September 30). The monetary policy strategy review: Some preliminary considerations. Speech by Christine Lagarde, president of the ECB, at the "ECB and its watchers xxi' conference, Frankfurt am main. European Central Bank.

ECB. (2022). *ECB takes further steps to incorporate climate change into its monetary policy operations* (Press Release). European Central Bank.
Elderson, F. (2018). *From mission to supervision: putting sustainable prosperity on the agenda of the Dutch central bank*. LSE sustainable finance leadership series.
Elderson, F. (2021, October 20). *Overcoming the tragedy of the horizon: Requiring banks to translate 2050 targets into milestones*. Keynote speech by Frank Elderson, Member of the Executive Board of the ECB and Vice-Chair of the Supervisory Board of the ECB, at the Financial Market Authority Supervisory Conference.
Elderson, F. (2022, September 28). *The ECB's monetary policy strategy: Delivering our mandate in all circumstances*. Keynote speech by Frank Elderson, Member of the Executive Board of the ECB and Vice-Chair of the Supervisory Board of the ECB, at the European Parliament conference on "Greening monetary policy in times of soaring inflation".
Faiella, I., & Lavecchia, L. (2020). The carbon content of Italian loans. *Journal of Sustainable Finance & Investment*, 1–19.
FSB. (2020). *The implications of climate change for financial stability*. Financial Stability Board.
Goodhart, C., et al. (2011). The changing role of central banks. *Financial History Review, 18*(2), 135–154.
HLEG. (2018). *Final report on financing a sustainable European economy*. Interim Report—High-Level Expert Group on Sustainable Finance.
IEA. (2018). *World energy investment 2018*. Publishing Paris. https://www.iea.org/publications/freepublications/publication/WEIO2014.pdf
IPCC. (2014). *Climate change 2014: Synthesis report* (Tech. Rep.). Intergovernmental Panel on Climate Change. http://www.ipcc.ch/report/ar5/syr/
IPCC. (2018). *Global warming of 1.5 degrees: Summary for policymakers* (Tech. Rep.). Intergovernmental Panel on Climate Change.
IPCC. (2021). *Ar6 climate change 2021: The physical science basis* (Tech. Rep.). Intergovernmental Panel on Climate Change.
IPCC. (2022). *Climate change 2022: Mitigation of climate change*. Intergovernmental Panel on Climate Change.
Krogstrup, S., & Oman, W. (2019). *Macroeconomic and financial policies for climate change mitigation: A review of the literature. International Monetary Fund, 19*. https://doi.org/10.5089/9781513511955.001
Matikainen, S., Campiglio, E., & Zenghelis, D. (2017). *The climate impact of quantitative easing* (Tech. Rep.). London School of Economics.
Mazzucato, M. (2015). The green entrepreneurial state. *The Politics of Green Transformations*, p. 134.

Mukherjee, K., & Ouattara, B. (2021). Climate and monetary policy: Do temperature shocks lead to inflationary pressures? *Climatic Change, 167*(3), 1–21.

NGFS. (2020). *Survey on monetary policy operations and climate change: Key lessons for further analyses.*

Olovsson, C. (2018). *Is climate change relevant for central banks?* Sveriges Riksbank Economic Commentaries, p. 13.

OMFIF. (2020). *Tackling climate change: The role of banking regulation and supervision.* Official Monetary and Financial Institution Forum & Luxembourg for Finance.

OMFIF. (2022). *Forging the path to international standards in sustainable finance.* Official Monetary and Financial Institution Forum & Luxembourg for Finance.

Polzin, F., & Sanders, M. (2020). How to finance the transition to low-carbon energy in Europe? *Energy Policy, 147*, 111863.

Sachs, J. D., Woo, W. T., Yoshino, N., & Taghizadeh-Hesary, F. (2019). Importance of green finance for achieving sustainable development goals and energy security. In *Handbook of green finance* (pp. 3–12). Springer.

Schnabel, I. (2020). When markets fail—The need for collective action in tackling climate change speech by Isabel Schnabel, member of the executive board of the ECB at the European sustainable finance summit, Frankfurt am Main. European Central Bank.

Schoenmaker, D. (2021). Greening monetary policy. *Climate Policy*, 1–12

Steuer, S., & Tröger, T. H. (2022). The role of disclosure in green finance. *Journal of Financial Regulation*, *8*(1), 1–50.

Stiglitz, J. E., Stern, N., Duan, M., Edenhofer, O., Giraud, G., Heal, G. M., Lèbre la Rovere, E., Morris, A., Moyer, E., Pangestu, M., Shukla, P. R., Sokona, Y., & Winkler, H. (2017). *Report of the high-level commission on carbon prices*. International Bank for Reconstruction and Development and International Development Association/The World Bank.

TCFD. (2017). *Final report: Recommendations of the task force on climate-related financial disclosures*. Task Force on Climate-related Financial Disclosures. https://www.fsb-tcfd.org/publications/final-recommendations-report

van't Klooster, J., & Fontan, C. (2020). The myth of market neutrality: A comparative study of the European Central Bank's and the Swiss National Bank's corporate security purchases. *New Political Economy*, *25*(6), 865–879.

Weidmann, J. (2019). Climate change and central banks. *Welcome address at the Deutsche Bundesbank's second financial market conference* (Vol. 29). *Frankfurt am Main.*

Weidmann, J. (2020, November 19). Central banks cannot solve climate change on their own. *Guest contribution by Jens Weidmann published in Financial Times.*

Index

A
Access to credit, 9, 19
Access to finance, 5
Accounting-reporting scandals, 99
Asymmetric information, 7, 15

B
bandwagon effect, 180, 181
Bank's ESG and lending relationships, 108
Banks' ESG value-relevance, 102, 104, 107, 116
Black Lives Matter, 98

C
capital requirements, 182, 183, 185, 188–190
central banks, 174–179, 186, 187
Charitable foundation, 2
climate change, 145, 147–149, 152, 154–156, 158, 162, 164, 165, 173–175, 177, 179, 185–188
climate finance, 148–151, 159, 162, 164, 165, 176, 177, 184
climate-related financial policies, 150, 166, 178, 184
climate-related financial policy index (CRFPI), 178
climate-related physical risks, 186, 188
climate-related risks, 153, 158, 182, 184, 189
climate-related transition risks, 175, 186
Climate risks, 97
Climate transition risk, 134–136
clusters, 180
Collateral, 9
Common good, 2, 3, 5, 16, 17, 19
Community development banks, 2, 4
Cooperative bank, 3, 4, 13
Corporate governance, 74, 78
Corporate Social Responsibility (CSR), 47–51, 53–61, 65, 66, 71, 74, 78, 79, 83, 85
credit allocation measures, 178

C. Gaganis et al. (eds.), *Sustainable Finance and ESG*, Palgrave Macmillan Studies in Banking and Financial Institutions, https://doi.org/10.1007/978-3-031-24283-0

Credit model, 128
Credit Rating Agencies (CRAs), 132, 133, 136–140
Credit rationing, 9, 10
Credit risk, 94, 96, 99, 100, 127–134, 137, 138, 140
Creditworthiness, 10, 11
Crowdfunding platform, 2, 19
CSR pyramid, 51, 60
Customer loyalty, 48, 49, 52–54, 56–60, 65
Customer satisfaction, 49, 51–60, 65, 66

D
Data Envelopment Analysis (DEA), 83
DealScan, 102, 111
Default risk, 129, 130
disclosure requirements, 146, 147, 150, 154, 156, 159, 161, 162, 179, 184, 188, 191
"Do well by doing good", 94

E
Employee treatment, 115
Environmental factor, 97, 115
Environmental, Social and Governance (ESG), 27–35, 37–40, 71–76, 79, 80, 84, 85, 91–98, 100–108, 110–114, 125–134, 136–140
ESG creates Long-term benefits, 95
ESG criteria, 191
ESG disclosure, 156, 161
Ethical banks, 15
Ethics in banking, 4
European Securities and Market Authority (ESMA), 132

F
Financial cooperative, 2
Financial sacrifice, 14–17
financial stability, 149, 153, 174–176, 186, 187, 189, 191
Financial Stability Board (FSB), 132, 151, 153
financial supervisors and regulators, 178
Fitch, 137–139

G
G20 countries, 150, 156, 160, 161
Gender diversity, 100
Global warming, 97
Governance factor, 99, 115
green bonds, 163–166, 177, 178, 187
green financial guidelines, 160, 161, 178
green monetary policy, 174, 186, 187
green regulations, 147, 154, 187
green taxonomy, 147, 154, 157, 158, 189
greenwashing, 101, 103, 162, 163, 166, 188

H
Hard information, 11, 18
High-level expert group on sustainable finance (HLEG), 156
"Hold up", 3

I
Impact investing, 2, 73
Interest rate, 14–19

L
Lending technology, 11

M
macroprudential policy, 174, 176, 184, 185
Management of financial institutions, 5
mandate, 175, 186, 187, 191
Market share, 6–8
Mean-variance model, 80, 84
#MeToo, 98, 99
Microbank, 3
Microcredit, 19
Microfinance institution, 2, 3, 9
microprudential policy, 174, 176, 182, 183
Moody's, 132, 133, 138, 139
Multicriteria Decision Analysis (MCDA), 83, 84
Mutual funds, 72, 74, 79–81, 83–85

N
Network for Greening the Financial System (NGFS), 147, 151, 154

O
Ohlson model (1995), 104

P
Paris Agreement, 145–148, 152, 154, 159, 163, 165, 173, 177, 178, 188, 189
Paris Climate Agreement, 92, 98
Physical risk, 97, 134–136
policy adoption, 174, 179, 180
policy diffusion, 174
Portfolio optimization, 72
Predatory lending, 4
Preferential loans, 17
price stability, 187
Principles of Responsible Investment (PRI), 92
prudential policies, 178, 184

R
Reciprocity, 15, 16
Regulatory framework, 18, 19
RepRisk, 102–104, 108
risk management, 134, 135, 139, 152, 154, 156, 158, 161, 163, 183, 188, 191

S
Savings bank, 3, 4, 18
Sea-level rise (SLR), 98, 115
Simple intermediation, 7, 12, 19
Social Banks (SBs), 1–19
Social capital, 99, 115
Social enterprise, 8–12, 17
Social factor, 98, 115
Social finance, 2, 5, 19
Socially responsible banks, 96, 101
Socially Responsible Investment (SRI), 71–76, 79–81, 83–85, 91
Social mission, 2, 4, 10, 13
Social premium, 14, 15, 18
Social project, 14, 18
Social screening, 11, 12
Soft information, 11
Speculation, 4, 5
Stakeholder theory, 50, 51, 55, 60, 61
Standard & Poor's (S&P's), 133–138
Sustainability-linked loans, 93
Sustainable banking, 28, 37
Sustainable Banking Network (SBN), 151, 153, 186
Sustainable finance, 37
sustainable finance landscape, 150, 152
sustainable investing, 149

T

Task Force on Climate-Related Financial Disclosures (TCFD), 132, 146, 153, 155, 156, 159
Task Force on Nature-related Financial Disclosures (TFND), 153
Transition risks, 97, 98
Transparency, 4, 12, 19
Triple bottom-line approach, 51
2008 Apex Oil settlement, 98, 115

U

UN's Sustainable Development Goals, 92
United Nations Sustainable Development Goals (SDGs), 145, 152
Usury, 2, 3

W

"Walk the ESG talk", 101
waves of adoption, 180, 181

MIX
Papier aus verantwortungsvollen Quellen
Paper from responsible sources
FSC® C105338

Printed by Books on Demand, Germany